ROYAL
OBSERVATORY
GREENWICH

2019 GUIDE
to the
NIGHT SKY

Storm Dunlop and Wil Tirion

FIREFLY BOOKS

A FIREFLY BOOK

Published by Firefly Books Ltd. 2018

First printing

Publisher Cataloging-in-Publication Data (U.S.)

Library of Congress Control Number: 2018939510

Library and Archives Canada Cataloguing in Publication

Dunlop, Storm, author
    2019 guide to the night sky / Storm Dunlop and Wil Tirion.
Includes bibliographical references.
ISBN 978-0-228-10105-5 (softcover)
    1. Astronomy--Observers' manuals. 2. Astronomy--Amateurs' manuals. 3. Astronomy--Popular works. I. Tirion, Wil, author II. Title. III. Title: Guide to the night sky. IV. Title: Two thousand nineteen guide to the night sky.
QB63.D86 2018   523   C2018-901917-4

Published in the United States by
Firefly Books (U.S.) Inc.
P.O. Box 1338, Ellicott Station
Buffalo, New York 14205

Published in Canada by
Firefly Books Ltd.
50 Staples Avenue, Unit 1
Richmond Hill, Ontario L4B 0A7

Printed in China by RR Donnelley APS

First published by Collins
An imprint of HarperCollins Publishers
Westerhill Road
Bishopbriggs
Glasgow G64 2QT

In association with
Royal Museums Greenwich, the group name for the National Maritime Museum, Royal Observatory Greenwich, Queen's House and *Cutty Sark*
www.rmg.co.uk

The contents of this publication are believed correct at the time of printing. Nevertheless the publisher can accept no responsibility for errors or omissions, changes in the detail given or for any expense or loss thereby caused.

The publisher does not warrant that any website mentioned in this title will be provided uninterrupted, that any website will be error free, that defects will be corrected, or that the website or the server that makes it available are free of viruses or bugs. For full terms and conditions please refer to the site terms provided on the website.

MIX
Paper from
responsible sources
FSC
www.fsc.org   **FSC™ C007454**

This book is produced from independently certified FSC™ paper to ensure responsible forest management.

For more information visit: www.harpercollins.co.uk/green

Contents

Introduction

The aim of this Guide is to help people find their way around the night sky, by showing how the stars that are visible change from month to month and by including details of various events that occur throughout the year. The objects and events described may be observed with the naked eye, or nothing more complicated than a pair of binoculars.

The conditions for observing naturally vary over the course of the year. During the summer, twilight may persist throughout the night and make it difficult to see the faintest stars. There are three recognized stages of twilight: civil twilight, when the Sun is less than 6° below the horizon; nautical twilight, when the Sun is between 6° and 12° below the horizon; and astronomical twilight, when the Sun is between 12° and 18° below the horizon. Full darkness occurs only when the Sun is more than 18° below the horizon. During nautical twilight, only the very brightest stars are visible. During astronomical twilight, the faintest stars visible to the naked eye may be seen directly overhead, but are lost at lower altitudes. As the diagram shows, during most of June full darkness never occurs at the latitude of Vancouver, BC. Slightly farther south (at Seattle, for example) it is truly dark for about two hours, and for somewhere like Houston, TX, there are at least six hours of darkness.

Another factor that affects the visibility of objects is the amount of moonlight in the sky. At Full Moon, it may be very difficult to see some of the fainter stars and objects, and even when the Moon is at a smaller phase it may seriously interfere with visibility if it is near the stars or planets in which you are interested. A full lunar calendar is given for each month and may be used to see when nights are likely to be darkest and best for observation.

The celestial sphere

All the objects in the sky (including the Sun, Moon and stars) appear to lie at some indeterminate distance on a large sphere, centered on the Earth. This *celestial sphere* has various reference points and features that are related to those of the Earth. If the Earth's rotational axis is extended, for example, it points to the North and South Celestial Poles, which are thus in line with the North and South Poles on Earth. Similarly, the *celestial equator* lies in the same plane as the Earth's equator, and divides the sky into northern and

The duration of twilight throughout the year at Vancouver and Houston.

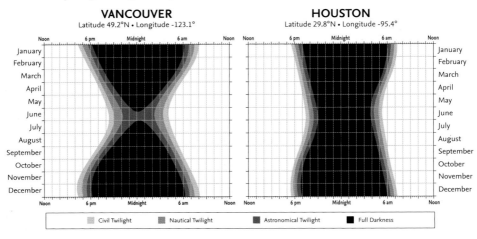

southern hemispheres. Because this Guide is written for use in North America, the area of the sky that it describes includes the whole of the northern celestial hemisphere and those portions of the southern that become visible at different times of the year. Stars in the far south, however, remain invisible throughout the year, and are not included.

It is useful to know some of the special terms for various parts of the sky. As seen by an observer, half of the celestial sphere is invisible, below the horizon. The point directly overhead is known as the **zenith**, and the (invisible) one below one's feet as the **nadir**. The line running from the north point on the horizon, up through the zenith and then down to the south point is the **meridian**. This is an important invisible line in the sky because objects are highest in the sky, and thus easiest to see, when they cross the meridian in the south. Objects are said to **transit** when they cross this line in the sky.

In this book, reference is frequently made in the text and in the diagrams to the standard compass points around the horizon. The position of any object in the sky may be described by its **altitude** (measured in degrees above the horizon), and its **azimuth** (measured in degrees from north 0°, through east 90°, south 180° and west 270°). Experienced amateurs and professional astronomers also use another system of specifying locations on the celestial sphere, but that need not concern us here, where the simpler method will suffice.

The celestial sphere appears to rotate about an invisible axis, running between the North and South Celestial Poles. The location (i.e., the altitude) of the Celestial Poles depends entirely on the observer's position on Earth or, more specifically, their latitude. The charts in this book are produced for the latitude of 40°N, so the North Celestial Pole (NCP) is 40° above the northern horizon. The fact that the NCP is fixed relative to the horizon means that all the stars within 40° of the pole are always above the horizon and may, therefore, always be seen at night, regardless of the time of year. This northern circumpolar region is an ideal place to begin learning the sky, and ways to identify the circumpolar stars and constellations will be described shortly.

The ecliptic and the zodiac

Another important line on the celestial sphere is the Sun's apparent path against the background stars – in reality the result

Measuring altitude and azimuth on the celestial sphere.

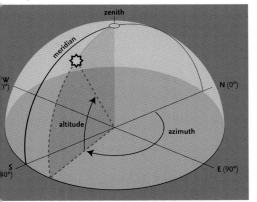

The altitude of the North Celestial Pole equals the observer's latitude.

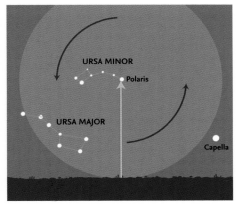

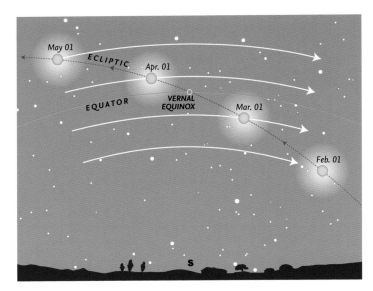

The Sun crossing the celestial equator in spring.

of the Earth's orbit around the Sun. This is known as the **ecliptic**. The point where the Sun, apparently moving along the ecliptic, crosses the celestial equator from south to north is known as the vernal (or spring) equinox, which occurs on March 20. At this time (and at the autumnal equinox, on September 22 or 23, when the Sun crosses the celestial equator from north to south) day and night are almost exactly equal in length. (There is a slight difference, but that need not concern us here.) The vernal equinox is currently located in the constellation of Pisces, and is important in astronomy because it defines the zero point for a system of celestial coordinates, which is, however, not used in this Guide.

The Moon and planets are to be found in a band of sky that extends 8° on either side of the ecliptic. This is because the orbits of the Moon and planets are inclined at various angles to the ecliptic (i.e., to the plane of the Earth's orbit). This band of sky is known as the zodiac and, when originally devised, consisted of 12 **constellations**, all of which were considered to be exactly 30° wide. When the constellation boundaries were formally established by the International Astronomical Union in 1930, the exact extent of most constellations was altered and, nowadays, the ecliptic passes through 13 constellations. Because of the boundary changes, the Moon and planets may actually pass through several other constellations that are adjacent to the original 12.

The constellations

Since ancient times, the celestial sphere has been divided into various constellations, most dating back to antiquity and usually associated with certain myths or legendary people and animals. Nowadays, the boundaries of the constellations have been fixed by international agreement and their names (in Latin) are largely derived from Greek or Roman originals. Some of the names of the most prominent stars are of Greek or Roman origin, but many are derived from Arabic names. Many bright stars have no individual names and, for many years, stars were identified by terms such as "the star in Hercules' right foot." A more sensible scheme was introduced by the German astronomer Johannes Bayer in the early 17th century. Following his scheme – which is still used today – most of the brightest stars are identified by a Greek letter followed by the genitive form of the constellation's Latin name. An example is the Pole Star, also known as Polaris and α Ursae Minoris (abbreviated α UMi). The Greek alphabet is shown on page 94 with a list of all the constellations that may be seen from latitude 40°N, together with abbreviations, their genitive forms and English names on page 93. Other naming schemes exist for fainter stars, but are not used in this book.

Asterisms

Apart from the constellations (88 of which cover the whole sky), certain groups of stars, which may form a part of a larger constellation or cross several constellations, are readily recognizable and have been given individual names. These groups are known as *asterisms*, and the most famous (and well-known) is the "Big Dipper," the common name for the seven brightest stars in the constellation of Ursa Major, the Great Bear. The names and details of some asterisms mentioned in this book are given in the list on page 94.

Magnitudes

The brightness of a star, planet or other body is frequently given in magnitudes (mag.). This is a mathematically defined scale where larger numbers indicate a fainter object. The scale extends beyond the zero point to negative numbers for very bright objects. (Sirius, the brightest star in the sky is mag. -1.4.) Most observers are able to see stars to about mag. 6, under very clear skies.

The Moon

Although the daily rotation of the Earth carries the sky from east to west, the Moon gradually moves eastwards by approximately its diameter (about half a degree) in an hour. Normally, in its orbit around the Earth, the Moon passes above or below the direct line between Earth and Sun (at New Moon) or outside the area obscured by the Earth's shadow (at Full Moon). Occasionally, however, the three bodies are more or less perfectly aligned to give an *eclipse*: a solar eclipse at New Moon or a lunar eclipse at Full Moon. Depending on the exact circumstances, a solar eclipse may be merely partial (when the Moon does not cover the whole of the Sun's disk), annular (when the Moon is too far from Earth in its orbit to appear large enough to hide the whole of the Sun), or total. Total and annular eclipses are visible from very restricted areas of the Earth, but partial eclipses are normally visible over a wider area.

Somewhat similarly, at a lunar eclipse, the Moon may pass through the outer zone of the Earth's shadow, the **penumbra** (in a penumbral eclipse, which is not generally perceptible to the naked eye), so that just part of the Moon is within the darkest part of the Earth's shadow, the **umbra** (in a partial eclipse); or completely within the umbra (in a total eclipse). Unlike solar eclipses, lunar eclipses are visible from large areas of the Earth.

Occasionally, as it moves across the sky, the Moon passes between the Earth and individual planets or distant stars, giving rise to an **occultation**. As with solar eclipses, such occultations are visible from restricted areas of the world.

The planets

Because the planets are always moving against the background stars, they are treated in some detail in the monthly pages and information is given when they are close to other planets, the Moon or any of five bright stars that lie near the ecliptic. Such events are known as **appulses** or, more frequently, as **conjunctions**. (There are technical differences in the way these terms are defined and should be used in astronomy, but these need not concern us here.) The positions of the planets are shown for every month on a special chart of the ecliptic.

The term conjunction is also used when a planet is either directly behind or in front of the Sun, as seen from Earth. (Under normal circumstances it will then be invisible.) The conditions of most favorable visibility depend on whether the planet is one of the two known

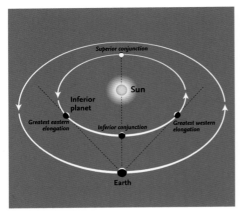

Inferior planet.

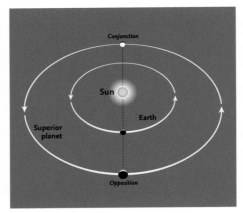

Superior planet.

as **inferior planets** (Mercury and Venus) or one of the three **superior planets** (Mars, Jupiter and Saturn) that are covered in detail. (Some details of the fainter superior planets, Uranus and Neptune, are included in this Guide, and special charts for both are given on pages 75 and 71.)

The inferior planets are most readily seen at eastern or western **elongation**, when their angular distance from the Sun is greatest. For superior planets, they are best seen at **opposition**, when they are directly opposite the Sun in the sky, and cross the meridian at local midnight.

It is often useful to be able to estimate angles on the sky, and approximate values may be obtained by holding one hand at arm's length. The various angles are shown in the diagram, together with the separations of the various stars in the Big Dipper.

Meteors

At some time or other, nearly everyone has seen a **meteor** – a "shooting star" – as it flashed across the sky. The particles that cause meteors – known technically as "meteoroids" – range in size from that of a grain of sand (or even smaller) to the size of a pea. On any night of the year there are occasional meteors, known as **sporadics**, that may travel in any direction. These occur at a rate that is normally between three and eight in an hour. Far more

important, however, are **meteor showers**, which occur at fixed periods of the year, when the Earth encounters a trail of particles left behind by a comet or, very occasionally, by a minor planet (asteroid). Meteors always appear to diverge from a single point on the sky, known as the **radiant**, and the radiants of major showers are shown on the charts. Meteors that come from a circular area 8° in diameter around the radiant are classed as belonging to the particular shower. All others that do not come from that area are sporadics (or, occasionally from another shower that is active at the same time). A list of the major meteor showers is given on page 17.

Although the positions of the various shower radiants are shown on the charts, looking directly at the radiant is not the most effective way of seeing meteors. They are most likely to be noticed if one is looking about 40°–45° away from the radiant position. (This is approximately two hand-spans as shown in the diagram for measuring angles.)

Other objects

Certain other objects may be seen with the naked eye under good conditions. Some were given names in antiquity – Praesepe is one example – but many are known by what are called "Messier numbers," the numbers in a catalogue of nebulous objects compiled by Charles Messier in the late 18th century.

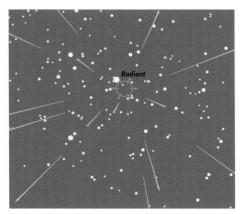

Meteor shower (showing the April Lyrid radiant).

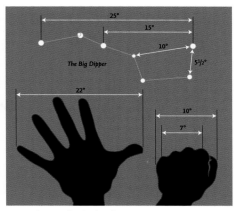

Measuring angles in the sky.

Some, such as the Andromeda Galaxy, M31, and the Orion Nebula, M42, may be seen by the naked eye, but all those given in the list will benefit from the use of binoculars. Apart from galaxies, such as M31, which contain thousands of millions of stars, there are also two types of cluster: open clusters, such as M45, the Pleiades, which may consist of a few dozen to some hundreds of stars; and globular clusters, such as M13 in Hercules, which are spherical concentrations of many thousands of stars. One or two gaseous nebulae, consisting of gas illuminated by stars within them, are also visible. The Orion Nebula, M42, is one, and is illuminated by the group of four stars, known as the Trapezium, which may be seen within it by using a good pair of binoculars.

Some interesting objects.

Messier / NGC	Name	Type	Constellation	Maps (months)
—	Hyades	open cluster	Taurus	Sep. – Apr.
—	Double Cluster	open cluster	Perseus	All year
—	Melotte 111 (Coma Cluster)	open cluster	Coma Berenices	Jan. – Aug.
M3	—	globular cluster	Canes Venatici	Jan. – Sep.
M4	—	globular cluster	Scorpius	May – Aug.
M8	Lagoon Nebula	gaseous nebula	Sagittarius	Jun. – Sep.
M11	Wild Duck Cluster	open cluster	Scutum	May – Oct.
M13	Hercules Cluster	globular cluster	Hercules	Feb. – Nov.
M15	—	globular cluster	Pegasus	Jun. – Dec.
M22	—	globular cluster	Sagittarius	Jun. – Sep.
M27	Dumbbell Nebula	planetary nebula	Vulpecula	May – Dec.
M31	Andromeda Galaxy	galaxy	Andromeda	All year
M35	—	open cluster	Gemini	Oct. – May
M42	Orion Nebula	gaseous nebula	Orion	Nov. – Mar.
M44	Praesepe	open cluster	Cancer	Nov. – Jun.
M45	Pleiades	open cluster	Taurus	Aug. – Apr.
M57	Ring Nebula	planetary nebula	Lyra	Apr. – Dec.
M67	—	open cluster	Cancer	Dec. – May
NGC 752	—	open cluster	Andromeda	Jul. – Mar.
NGC 3242	Ghost of Jupiter	planetary nebula	Hydra	Feb. – May

The Northern Circumpolar Constellations

The northern circumpolar stars are the key to starting to identify the constellations. For anyone in the northern hemisphere they are visible at any time of the year, and nearly everyone is familiar with the seven stars of the Big Dipper: an asterism that forms part of the large constellation of **Ursa Major** (the Great Bear).

Ursa Major

Because of the movement of the stars caused by the passage of the seasons, Ursa Major lies in different parts of the evening sky at different periods of the year. The diagram below shows its position for the four main seasons. The seven stars of the Big Dipper remain visible throughout the year anywhere north of latitude 40°N. Even at the latitude (40°N) for which the charts in this book are drawn, many of the stars in the southern portion of the constellation of Ursa Major are hidden below the horizon for part of the year or (particularly in late summer) cannot be seen late in the night.

Polaris and Ursa Minor

The two stars **Dubhe** and **Merak** (α and β Ursae

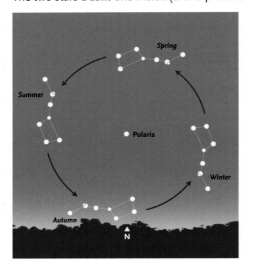

Majoris, respectively), farthest from the "tail" are known as the "Pointers." A line from Merak to Dubhe, extended about five times their separation, leads to the Pole Star, **Polaris**, or α Ursae Minoris. All the stars in the northern sky appear to rotate around it. There are five main stars in the constellation of **Ursa Minor**, and the two farthest from the Pole, **Kochab** and **Pherkad** (β and γ Ursae Minoris, respectively), are known as "The Guards."

Cassiopeia

On the opposite of the North Pole from Ursa Major lies **Cassiopeia**. It is highly distinctive, appearing as five stars forming a letter "W" or "M" depending on its orientation. Provided the sky is reasonably clear of clouds, you will nearly always be able to see either Ursa Major or Cassiopeia, and thus be able to orientate yourself on the sky.

To find Cassiopeia, start with **Alioth** (ε Ursae Majoris), the first star in the tail of the Great Bear. A line from this star extended through Polaris points directly toward γ Cassiopeiae, the central star of the five.

Cepheus

Although the constellation of **Cepheus** is fully circumpolar, it is not nearly as well-known as Ursa Major, Ursa Minor or Cassiopeia, partly because its stars are fainter. Its shape is rather like the end of a house with a pointed roof. The line from the Pointers through Polaris, if extended, leads to **Errai** (γ Cephei) at the "top" of the "roof." The brightest star, **Alderamin** (α Cephei) lies in the Milky Way region, at the "bottom right-hand corner" of the figure.

Draco

The constellation of **Draco** consists of a quadrilateral of stars, known as the "Head of Draco" (and also the "Lozenge"), and a long chain of stars forming the neck and body of the dragon. To find the Head of Draco, locate the two stars **Phecda** and **Megrez** (γ and δ

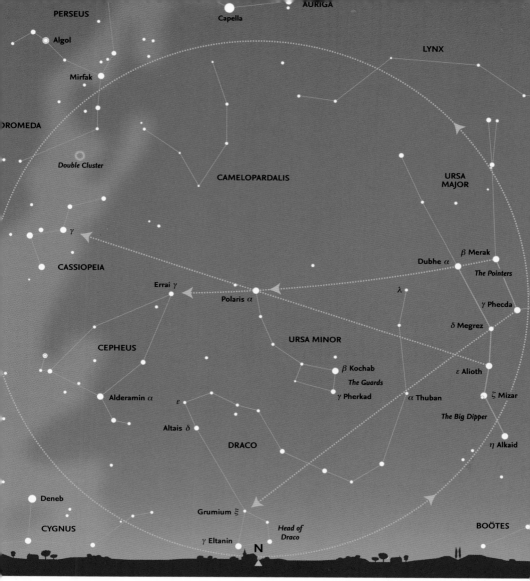

PERSEUS
Capella
AURIGA
Algol
LYNX
Mirfak
ROMEDA
Double Cluster
CAMELOPARDALIS
URSA MAJOR
γ
CASSIOPEIA
β Merak
Dubhe α
The Pointers
Errai γ
λ
γ Phecda
Polaris α
δ Megrez
CEPHEUS
URSA MINOR
β Kochab
ε Alioth
The Guards
Alderamin α
γ Pherkad
ε
α Thuban
ζ Mizar
Altais δ
The Big Dipper
DRACO
η Alkaid
Deneb
Grumium ξ
BOÖTES
CYGNUS
Head of Draco
γ Eltanin
N

The stars and constellations inside the circle are always above the horizon, seen from latitude 40°N.

Ursae Majoris) in the Big Dipper, opposite the Pointers. Extend a line from Phecda through Megrez by about eight times their separation, right across the sky below the Guards in Ursa Minor, ending at **Grumium** (ξ Draconis) at one corner of the quadrilateral. The brightest star, **Eltanin** (γ Draconis) lies farther to the south.

From the head of Draco, the constellation first runs northeast to **Altais** (δ Draconis) and ε Draconis, then doubles back southwards before winding its way through **Thuban** (α Draconis) before ending at λ Draconis between the Pointers and Polaris.

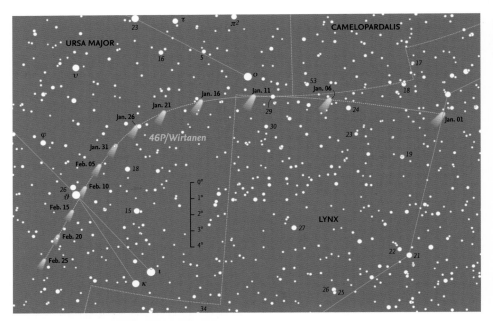

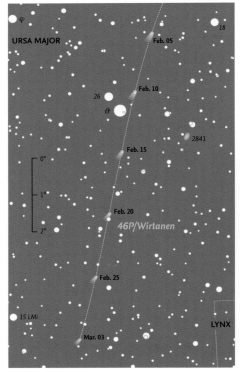

The path of comet 46P/Wirtanen from January 1 to March 3, 2019.

The chart above shows stars down to magnitude 8.5 and the chart on the left shows stars to magnitude 10.0.

Comet C/2014 Q2 Lovejoy, which reached naked-eye visibility, photographed on December 20, 2014, when close to κ Columbae, by Damian Peach.

Comets and the Moon

Comet C/2006 P1 McNaught, imaged on January 20, 2007, from Lawlers Gold Mine, Western Australia (Photographer: Sjbmgrtl).

Comets

Although comets may occasionally become very striking objects in the sky, their occurrence and particularly the existence or length of any tail and their overall magnitude are notoriously difficult to predict. Naturally, it is only possible to predict the return of periodic comets (whose names have the prefix "P"). Many comets appear unexpectedly (these have names with the prefix "C"). Bright, readily visible comets such as C/1995 Y1 Hyakutake & C/1995 O1 Hale-Bopp or C/2006 P1 McNaught (sometimes known as the Great Comet of 2007) are rare. (Comet Hale-Bopp, in particular, was visible for a record 18 months and was a prominent object in northern skies. Comet McNaught was notable for its multiple tail structure. Most periodic comets are faint and only a very small number ever become bright enough to be readily visible with the naked eye or with binoculars. One comet, 46P/Wirtanen, was expected to become visible in binoculars in October 2018. It is well placed in the northern sky in the first few months of 2019. The accompanying charts show its path at its brightest during early 2019 until March, when it is expected to fade below magnitude 10.

The Moon

The monthly pages include diagrams showing the phase of the Moon for every day of the month, and also indicate the day in the **lunation** (or **age** of the Moon), which begins at New Moon. Although the main features of the surface – the light highlands and the dark maria (seas) – may be seen with the naked eye, far more features may be detected with the use of binoculars or any telescope. The many craters are best seen when they are close to the **terminator** (the boundary between the illuminated and the non-illuminated areas of the surface), when the Sun rises or sets over any particular region of the Moon and the crater walls or central peaks cast strong shadows. Most features become difficult to see at Full Moon, although this is the best time to see the bright ray systems surrounding certain craters. Accompanying the Moon map on the following pages is a list of prominent features, including the days in the lunation when features are normally close to the terminator and thus easiest to see. A few bright features such as Linné and Proclus, visible when well illuminated, are also listed. One feature, Rupes Recta (the Straight Wall) is readily visible only when it casts a shadow with light from the east, appearing as a light line when illuminated from the opposite direction.

The dates of visibility vary slightly through the effects of **libration**. Because the Moon's orbit is inclined to the Earth's equator and also because it moves in an ellipse, the Moon appears to rock slightly from side to side (and nod up and down). Features near the **limb** (the edge of the Moon) may vary considerably in their location and visibility. (This is easily noticeable with Mare Crisium and the craters Tycho and Plato.) Another effect is that at crescent phases before and after New Moon, the normally non-illuminated portion of the Moon receives a certain amount of light, reflected from the Earth. This **Earthshine** may enable certain bright features (such as Aristarchus, Kepler and Copernicus) to be detected even though they are not illuminated by sunlight.

Map of the Moon

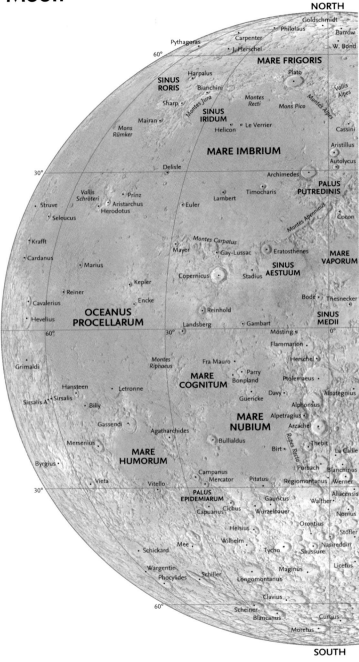

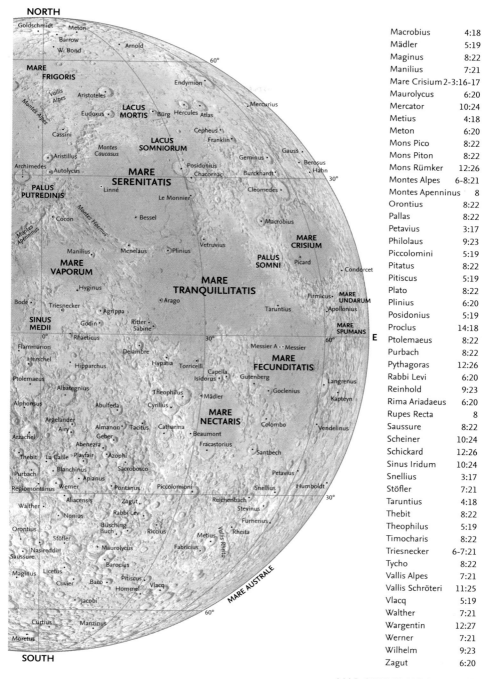

NORTH

SOUTH

Macrobius 4:18
Mädler 5:19
Maginus 8:22
Manilius 7:21
Mare Crisium 2-3:16-17
Maurolycus 6:20
Mercator 10:24
Metius 4:18
Meton 6:20
Mons Pico 8:22
Mons Piton 8:22
Mons Rümker 12:26
Montes Alpes 6-8:21
Montes Apenninus 8
Orontius 8:22
Pallas 8:22
Petavius 3:17
Philolaus 9:23
Piccolomini 5:19
Pitatus 8:22
Pitiscus 5:19
Plato 8:22
Plinius 6:20
Posidonius 5:19
Proclus 14:18
Ptolemaeus 8:22
Purbach 8:22
Pythagoras 12:26
Rabbi Levi 6:20
Reinhold 9:23
Rima Ariadaeus 6:20
Rupes Recta 8
Saussure 8:22
Scheiner 10:24
Schickard 12:26
Sinus Iridum 10:24
Snellius 3:17
Stöfler 7:21
Taruntius 4:18
Thebit 8:22
Theophilus 5:19
Timocharis 8:22
Triesnecker 6-7:21
Tycho 8:22
Vallis Alpes 7:21
Vallis Schröteri 11:25
Vlacq 5:19
Walther 7:21
Wargentin 12:27
Werner 7:21
Wilhelm 9:23
Zagut 6:20

Introduction to the Month-by-Month Guide

The monthly charts

The pages devoted to each month contain a pair of charts showing the appearance of the night sky, looking north and looking south. The charts (as with all the charts in this book) are drawn for the latitude of 40°N, so observers farther north will see slightly more of the sky on the northern horizon, and slightly less on the southern. These areas are, of course, those most likely to be affected by poor observing conditions caused by haze, mist or smoke. In addition, stars close to the horizon are always dimmed by atmospheric absorption, so sometimes the faintest stars marked on the charts may not be visible.

The three times shown for each chart require a little explanation. The charts are drawn to show the appearance at 11 p.m. for the 1st of each month. The same appearance will apply an hour earlier (10 p.m.) on the 15th, and yet another hour earlier (9 p.m.) at the end of the month (shown as the 1st of the following month). These times are local time, and apply in all time zones. Where Daylight Saving Time (DST) is used, in 2019 it is introduced on March 10, and ends on November 3. When used, both local time and DST are shown on the charts. Times of specific events are shown in the 24-hour clock of Universal Time (UT) used by astronomers worldwide, and the correct zone time (and DST, where appropriate) may be found from the details inside the front cover.

The charts may be used for earlier or later times during the night. To observe two hours earlier, use the charts for the preceding month; for two hours later, the charts for the next month.

Meteors

Details of specific meteor showers are given in the months when they come to maximum, regardless of whether they begin or end in other months. Note that not all the respective radiants are marked on the charts for that

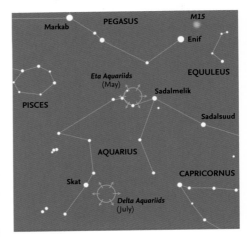

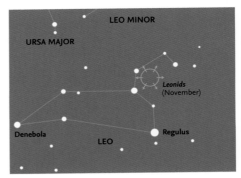

particular month, because the radiants may be below the horizon, or lie in constellations that are not readily visible during the month of maximum. For this reason, special charts for the Eta and Delta Aquariids (May and July, respectively) and the Leonids (November) are shown above. As explained earlier, however, meteors from such showers may still be seen, because the most effective region for seeing meteors is some 40°–45° away from the radiant, and that area of sky may well be above the horizon. A table of the best meteor showers visible during the year is also given here.

Meteors that are brighter than magnitude -4 (approximately the maximum magnitude reached by Venus) are known as **fireballs** or

Shower	Dates of activity 2019	Date of maximum 2019	Possible hourly rate
Quadrantids	January 1–10	January 3–4	120
April Lyrids	April 16–25	April 21–22	18
Eta Aquariids	April 19 to May 26	May 6–7	55
Alpha Capricornids	July 11 to August 10	July 26 to August 1	5
Perseids	July 13 to August 26	August 11–12	100
Delta Aquariids	July 21 to August 23	July 29–30	< 20
Alpha Aurigids	August to October	August 28 & September 15	10
Southern Taurids	September 23 to November 19	October 28–29	< 5
Orionids	September 23 to November 27	October 21–22	25
Northern Taurids	October 19 to December 10	November 10–11	< 5
Leonids	November 5–30	November 17–18	< 15
Geminids	December 4–16	December 13–14	100+
Ursids	December 17–23	December 21–22	< 10

bolides. Examples are shown on page 23 and 63. Fireballs sometimes cause sonic booms that may be heard some time after the meteor is seen.

The photographs
As an aid to identification – especially as some people find it difficult to relate charts to the actual stars they see in the sky – one or more photographs of constellations visible in certain specific months are included. It should be noted, however, that because of the limitations of the photographic and printing processes, and the differences between the sensitivity of different individuals to faint starlight (especially in their ability to detect different colors), and the degree to which they have become adapted to the dark, the apparent brightness of stars in the photographs will not necessarily precisely match that seen by any one observer.

The Moon calendar
The Moon calendar is largely self-explanatory. It shows the phase of the Moon for every day of the month, with the exact times (in Universal Time) of New Moon, First Quarter, Full Moon and Last Quarter. Because the times are calculated from the Moon's actual orbital parameters, some of the times shown will, naturally, fall during daylight, but any difference is too small to affect the appearance of the Moon on that date. Also shown is the age of the Moon (the day in the **lunation**), beginning at New Moon, which may be used to determine the best time for observation of specific lunar features.

The Moon
The section on the Moon includes details of any lunar or solar eclipses that may occur during the month (visible from anywhere on Earth). Similar information is given about any important occultations. Mainly, however, this section summarizes when the Moon passes close to planets or the five prominent stars close to the ecliptic. The dates when the Moon is closest to the Earth (at **perigee**) and farthest from it (at **apogee**) are shown in the monthly calendars, and only mentioned here when they are particularly significant, such as the nearest and farthest during the year.

The planets and minor planets
Brief details are given of the location, movement and brightness of the planets from Mercury to Saturn throughout the month. None of the planets can, of course, be seen when they are close to the Sun, so such periods are generally noted. All of the planets may sometimes lie on the opposite side of the Sun to the Earth (at superior conjunction), but in the case of the inferior planets, Mercury and Venus, they may also pass between the Earth and the Sun (at inferior conjunction) and are normally invisible for a period of time,

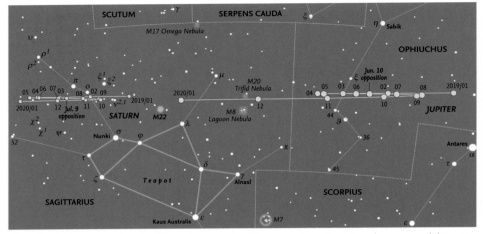

The paths of Jupiter and Saturn in 2019. Jupiter comes to opposition on June 10; Saturn almost a month later, on July 9. Background stars are shown down to magnitude 6.5.

the length of which varies from conjunction to conjunction. Those two planets are normally easiest to see around either eastern or western elongation, in the evening or morning sky, respectively. Not every elongation is favorable, so although every elongation is listed, only those where observing conditions are favorable are shown in the individual diagrams of events.

The dates at which the superior planets reverse their motion (from direct motion to **retrograde**, and retrograde to direct) and of opposition (when a planet generally reaches its maximum brightness) are given. Some planets, especially distant Saturn, may spend most or all of the year in a single constellation. Jupiter and Saturn are normally easiest to see around opposition, which occurs every year. Mars, by contrast, moves relatively rapidly against the background stars and in some years never comes to opposition.

Uranus is not always included in the monthly details because it is generally at the limit of naked-eye visibility (magnitude 5.7–5.9), although bright enough to be visible in binoculars, or even with the naked eye under exceptionally dark skies. Its path in 2019 is shown on the special chart in October. It comes to opposition on October 28 in the constellation of Aries. New Moon occurs that day, so the planet, at magnitude 5.7 should be detectable reasonably easily. It is at the same magnitude for an extended period of the year (from August to December) and should be visible when free from interference by moonlight.

Similar considerations apply to Neptune, although this is always fainter (magnitude 7.8–8.0 in 2019), but still visible in most binoculars. It reaches opposition at mag. 7.8 on September 10 in Aquarius. Full Moon occurs four days later, so Neptune will be difficult to detect for about a week after Full Moon. Again, Neptune's path in 2019 and its position at opposition are shown in a chart in September.

Charts for the three brightest minor planets that come to opposition in 2019 are shown in the relevant month: Pallas (mag. 7.9) on April 10, Ceres (mag.7.0) on May 28, and Vesta (mag. 6.5) on November 12.

The ecliptic charts

Although the ecliptic charts are primarily designed to show the positions and motions of the major planets, they also show the motion of the Sun during the month. The light-tinted area shows the area of the sky that is invisible during daylight, but the darker area gives an indication of which constellations are likely to be visible at some time of the night. The closer a planet is to the border between dark and light, the more difficult it will be to see in the twilight.

The monthly calendar

For each month, a calendar shows details of significant events, including when planets are close to one another in the sky, close to the Moon, or close to any one of five bright stars that are spaced along the ecliptic. The times shown are given in Universal Time (UT), always used by astronomers throughout the year. The tables given inside the front cover may be used to convert UT to that used in any given time zone and (during the summer) to DST.

The diagrams of interesting events

Each month, a number of diagrams show the appearance of the sky when certain events take place. However, the exact positions of celestial objects and their separations greatly depend on the observer's position on Earth. When the Moon is one of the objects involved, because it is relatively close to Earth, there may be very significant changes from one location to another. Close approaches between planets or between a planet and a star are less affected by changes of location, which may thus be ignored.

The diagrams showing the appearance of the sky are drawn for the latitude of 40°N and 90°W, northwest of Springfield, IL, so will be approximately correct for most of North America. However, for an observer farther north (say Vancouver), a planet or star listed as being north of the Moon will appear even farther north, whereas one south of the Moon will appear closer to it – or may even be hidden (occulted) by it. For an observer at a latitude of less than 40°N, there will be corresponding changes in the opposite direction: for a star or planet south of the Moon the separation will increase, and for one north of the Moon the separation will decrease. For example, on January 23, 2019, during the conjunction of Regulus (pages 24 and 25), the apparent position of the star is closer to the Moon seen from Vancouver, than from Houston.

Ideally, details should be calculated for each individual observer, but this is obviously impractical. In fact, positions and separations are actually calculated for a theoretical observer located at the center of the Earth.

So the details given regarding the positions of the various bodies should be used as a guide to their location. A similar situation arises with the times that are shown. These are calculated according to certain technical criteria, which need not concern us here. However, they do not necessarily indicate the exact time when two bodies are closest together. Similarly, dates and times are given, even if they fall in daylight, when the objects are likely to be completely invisible. However, such times do give an indication that the objects concerned will be in the same general area of the sky during both the preceding, and the following nights.

Key to the symbols used on the monthy star maps.

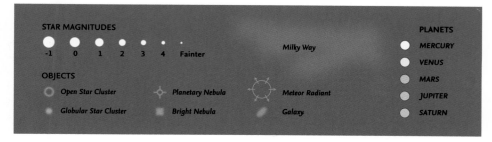

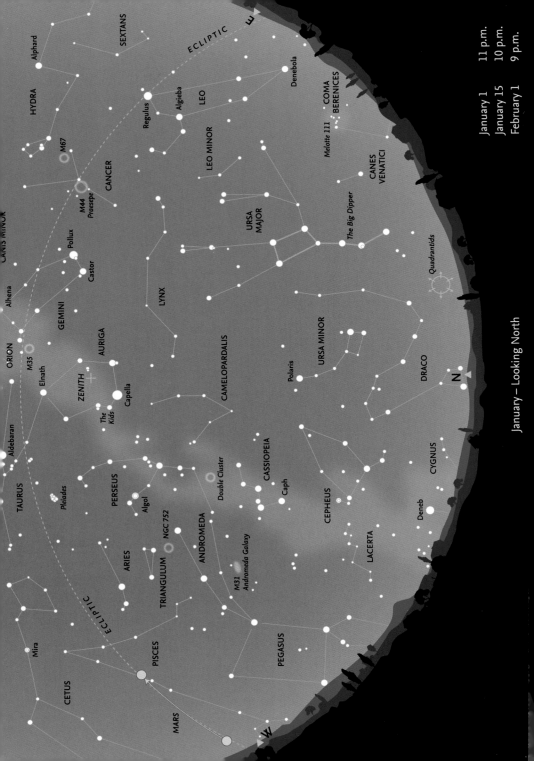

January – Looking North

January – Looking North

Most of the important circumpolar constellations are easy to see in the northern sky at this time of year. **Ursa Major** stands more-or-less vertically above the horizon in the northeast, with the zodiacal constellation of **Leo** rising in the east. To the north, the stars of **Ursa Minor** lie below **Polaris** (the Pole Star). The head of **Draco** is low on the northern horizon, but may be difficult to see unless observing conditions are good. Both **Cepheus** and **Cassiopeia** are readily visible in the northwest, and even the faint constellation of **Camelopardalis** is high enough in the sky for it to be easily visible.

Near the zenith is the constellation of **Auriga** (the Charioteer), with brilliant **Capella** (α Aurigae), directly overhead. Slightly to the west of Capella lies a small triangle of fainter stars, known as "The Kids." (Ancient mythological representations of Auriga show him carrying two young goats.) Together with the northernmost bright star in Taurus, **Elnath** (β Tauri), the body of Auriga forms a large pentagon on the sky, with The Kids lying on the western side. Farther down toward the west are the constellations of **Perseus** and **Andromeda**, and the Great Square of **Pegasus** is approaching the horizon.

Meteors

One of the strongest and most consistent meteor showers of the year occurs in January: the **Quadrantids**, which are visible January 1–10, with maximum on January 3–4. They are brilliant, bluish and yellowish-white meteors and fireballs (page 23), which at maximum may even reach a rate of 120 meteors per hour. At maximum, the Moon is a waning crescent, just before New Moon, so there will be little interference from moonlight. The parent object is minor planet 2003 EH$_1$.

The shower is named after the former constellation **Quadrans Muralis** (the Mural Quadrant), an early form of astronomical instrument. The Quadrantid meteor radiant is now within the northernmost part of **Boötes**, roughly halfway between θ Boötis and τ Herculis. This is very low on the northern horizon, lying approximately halfway between **Alioth**, the last star in the "handle" of the **Big Dipper**, and the head of **Draco**.

The constellation of Orion dominates the sky during this period of the year, and is a useful starting point for recognizing other constellations in the southern sky. Here, orange Betelgeuse, blue-white Rigel and the pinkish Orion Nebula are prominent. Orion can be found in the southern part of the sky (see next page).

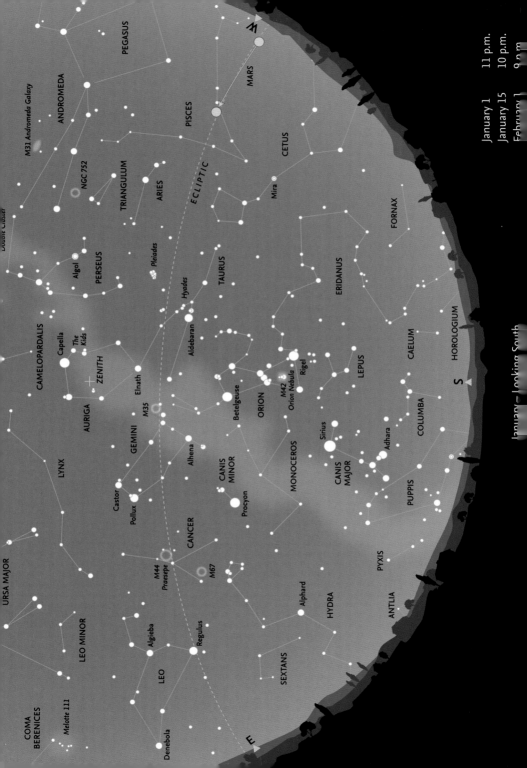

January — Looking South

January 1   11 p.m.
January 15   10 p.m.
February 1   9 p.m.

PEGASUS
ANDROMEDA
M31 Andromeda Galaxy
PISCES
MARS
CETUS
Mira
NGC 752
TRIANGULUM
ARIES
ECLIPTIC
FORNAX
Double Cluster
Algol
PERSEUS
Pleiades
TAURUS
ERIDANUS
CAMELOPARDALIS
Capella
The Kids
Hyades
Aldebaran
ZENITH
Elnath
M35
CAELUM
S
HOROLOGIUM
AURIGA
GEMINI
Rigel
M42
Orion Nebula
LEPUS
LYNX
Castor
Pollux
Alhena
Betelgeuse
ORION
Sirius
COLUMBA
CANIS
MINOR
MONOCEROS
Adhara
CANIS
MAJOR
PUPPIS
Procyon
URSA MAJOR
CANCER
PYXIS
LEO MINOR
Algieba
M44
Praesepe
M67
ANTLIA
COMA
BERENICES
Melotte 111
LEO
Regulus
Alphard
HYDRA
SEXTANS
Denebola
E

January – Looking South

At this time of year the southern sky is dominated by **Orion**. This is the most prominent constellation during the winter months, when it is visible at some time during the night. (A photograph of Orion appears on page 21.) It has a highly distinctive shape, with a line of three stars that form the "Belt." To most observers, the bright star at the northeastern corner of the constellation, **Betelgeuse** (α Orionis), shows a reddish tinge, in contrast to the brilliant bluish-white color of the bright star at the southwestern corner, **Rigel** (β Orionis). The three stars of the belt lie directly south of the celestial equator. A vertical line of three "stars" forms the "Sword" that hangs to the south of the Belt. With good viewing conditions, the central "star" appears as a hazy spot, even to the naked eye. This is actually the **Orion Nebula**. Binoculars will reveal the four stars of the **"Trapezium,"** which illuminate the nebula.

The line of Orion's Belt points up to the northwest toward **Taurus** (the Bull) and below orange-tinted **Aldebaran** (α Tauri). Close to Aldebaran, there is a conspicuous "V" of stars, pointing down to the southwest, called the **Hyades** cluster. (Despite appearances, Aldebaran is not part of the cluster.) Farther along, the same line from Orion

A late Quadrantid fireball, photographed from Portmahomack, Ross-shire, Scotland on January 15, 2018 at 23:44 UT.

passes below a bright cluster of stars, the **Pleiades**, or Seven Sisters. Even the smallest pair of binoculars reveals this cluster to be a beautiful group of bluish-white stars. The two most conspicuous of the other stars in Taurus lie directly above Orion, and form an elongated triangle with Aldebaran. The northernmost, β Tauri, was once considered to be part of the constellation of Auriga.

The Moon's phases for January

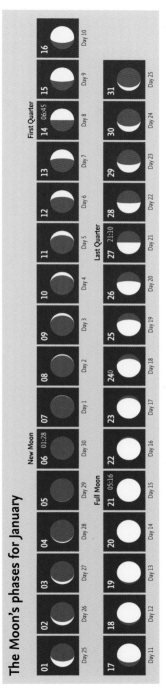

January – Moon and Planets

The Earth

The Earth reaches perihelion (the closest point to the Sun in its annual orbit) on January 3, 2019, at 05:20 Universal Time. Its distance is then 0.9833 AU (147,099,586 km).

The Moon

On January 3 the Moon is close to Jupiter in Scorpius. On January 6 at New Moon there is a partial solar eclipse, visible from the region of the northwestern Pacific and northeastern Asia. On January 17, the Moon is close to **Aldebaran** in **Taurus**. A total lunar eclipse occurs at Full Moon on January 21, visible from a wide area of the Pacific, including Australia and eastern Asia. The Moon passes close to **Regulus** in **Leo** on January 23.

The planets

Mercury is too close to the Sun to be readily visible this month. **Venus** reaches greatest elongation west (47°) on January 6, when its magnitude is -4.6. **Mars** moves across the constellation of **Pisces**, fading from mag. 0.5 to 0.9 over the month. **Jupiter** is fairly bright (mag. -1.9 to -1.8) and in **Ophiuchus**, seen only in the early morning. **Saturn** is in **Sagittarius**, too near the Sun to be visible. **Uranus** (mag. 5.8) is on the border of **Pisces** and **Aries**. **Neptune** (mag. 7.9) is in **Aquarius**, where it remains throughout the year.

The path of the Sun and the planets along the ecliptic in January.

Calendar for January

01–10		Quadrantid meteor shower
01	21:48	Venus 1.3°S of Moon
02	05:50	Saturn in conjunction with Sun
03–04		Quadrantid shower maximum
03	05:20	Earth at Perihelion (147,099,586 km = 0.9833 AU)
03	07:35	Jupiter 3.1°S of Moon
04	17:40	Mercury 2.7°S of Moon
06	01:28	New Moon
06	01:41	Partial solar eclipse (NW Pacific, NE Asia)
06	04:59	Venus at greatest elongation (47°W, mag. -4.6)
09	04:29	Moon at apogee (406,117 km)
12	19:47	Mars 5.3°N of Moon
14	06:45	First Quarter
15	21:00 *	Venus 7.9°N of Antares
17	18:20	Aldebaran 1.6°S of Moon
21	05:12	Total lunar eclipse (Pacific, Australia, E. Asia)
21	05:16	Full Moon
21	20:00	Moon at perigee (357,342 km)
22	15:10	Venus 2.4°N of Jupiter
23	01:41	Regulus 2.5°S of Moon
27	21:10	Last Quarter
30	02:46	Mercury superior conjunction
30	23:54	Jupiter 3°S of Moon
31	17:36	Venus 0.1°S of Moon

* These objects are close together for an extended period around this time.

Morning 6:30 a.m.

January 1–3 • The Moon passes Venus, Antares and Jupiter, early in the morning.

Evening 5:30 p.m.

January 17 • The Moon passes Aldebaran, high in the southeast.

Morning 6:30 a.m.

January 22 • Venus and Jupiter close together, in the company of Sabik and Antares.

Morning 6:30 a.m.

January 30 – February 1 • The Moon with Antares, Jupiter and Venus. Saturn is low in the southeast.

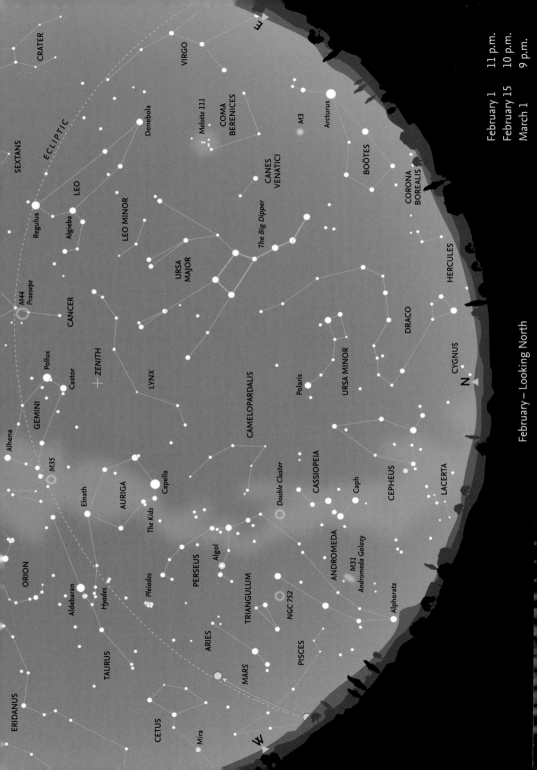

February – Looking North

February 1   11 p.m.
February 15   10 p.m.
March 1   9 p.m.

February – Looking North

The months of January and February are probably the best time for seeing the section of the Milky Way that runs in the northern and western sky from **Cygnus**, low on the northern horizon, through **Cassiopeia**, **Perseus** and **Auriga** and then down through **Gemini** and **Orion**. Although not as readily visible as the denser star clouds of the summer Milky Way, on a clear night so many stars may be seen that even a distinctive constellation such as **Cassiopeia** is not immediately obvious.

The head of **Draco** is now higher in the sky and easier to recognize. **Deneb**, (α Cygni), the brightest star in **Cygnus**, may just be visible almost due north at midnight, early in the month, if the sky is very clear and the horizon clear of obstacles. **Vega** (α Lyrae) in **Lyra** is so low that it is difficult to see, but may become visible later in the night. The constellation of **Boötes** – sometimes described as shaped like a kite, an ice-cream cone, or the letter "p" – with orange-tinted **Arcturus** (α Boötis), is beginning to clear the eastern horizon. Arcturus, at magnitude -0.05, is the brightest star in the northern hemisphere. The inconspicuous constellation of **Coma Berenices** is now well above the horizon in the east. The concentration of faint stars at the northwestern corner somewhat resembles a tiny, detached portion of the Milky Way. This is Melotte 111, an open star cluster (which is sometimes called the Coma Cluster, but must not be confused with the important Coma Cluster of galaxies, Abell 1656, mentioned on page 41).

On the other side of the sky, in the northwest, most of the constellation of **Andromeda** is still easily seen, although **Alpheratz** (α Andromedae), the star that forms the northeastern corner of the Great Square of Pegasus – even though it is actually part of Andromeda – is becoming close to the horizon and more difficult to detect. High overhead, at the zenith, try to make out the very faint constellation of

Lynx. It was introduced in 1687 by the famous astronomer Johannes Hevelius to fill the largely blank area between **Auriga**, **Gemini** and **Ursa Major**, and is reputed to be so named because one needed the eyes of a lynx to detect it.

A very large, and frequently ignored, open star cluster, Melotte 111, also known as the Coma Cluster, is readily visible in the eastern sky during February.

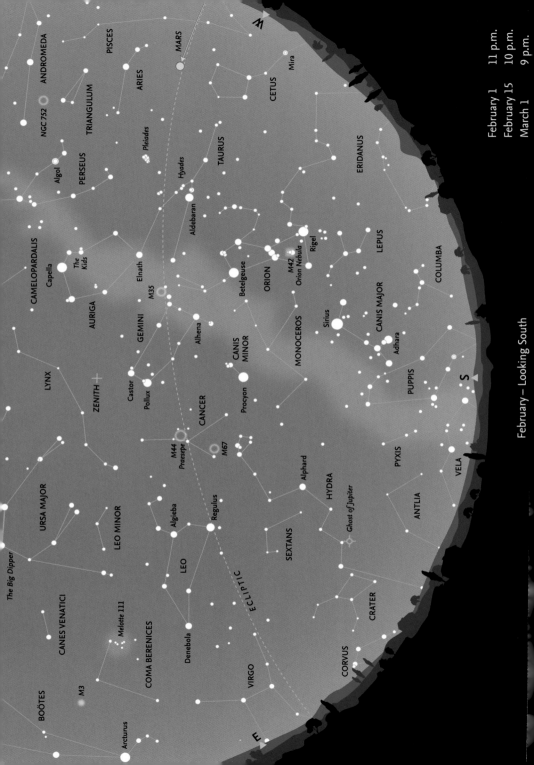

February – Looking South

February 1   11 p.m.
February 15   10 p.m.
March 1   9 p.m.

February – Looking South

Apart from Orion, the most prominent constellation visible this month is **Gemini**, with its two lines of stars running southwest towards **Orion**. Many people have difficulty in remembering which is which of the two stars **Castor** and **Pollux**. Think of them in alphabetical order: Castor (α Geminorum), the fainter star, is closer to the North Celestial Pole. Pollux (β Geminorum) is the brighter of the two, but is farther away from the Pole. Castor is remarkable because it is actually a multiple system, consisting of no less than six individual stars.

Using Orion's belt as a guide, it points down to the southeast towards **Sirius**, the brightest star in the sky (at magnitude -1.4) in the constellation of **Canis Major**, and the southern constellation of **Puppis**. Forming an equilateral triangle with **Betelgeuse** in **Orion** and **Sirius** in Canis Major is **Procyon**, the brightest star in the small constellation of **Canis Minor**. Between Canis Major and Canis Minor is the faint constellation of **Monoceros**, which actually straddles the Milky Way, which, although faint, has many clusters in this area. Directly east of Procyon is the highly distinctive asterism of six stars that form the "head" of **Hydra**, the largest of all 88 constellations, and which trails

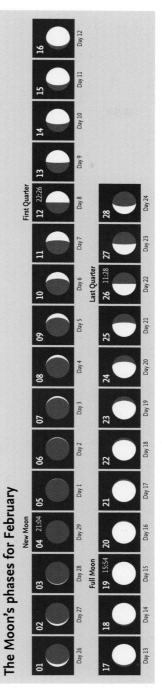

The constellation of Gemini. The two brightest stars are Castor and Pollux, visible on the left-hand side of the photograph.

such a long way across the sky that it is only in mid-March around midnight that the whole constellation becomes visible.

The Moon's phases for February

New Moon							
01	02	03	04 21:04	05 21:04	06	07	08
Day 26	Day 27	Day 28	Day 29	Day 1	Day 2	Day 3	Day 4

Full Moon appears near Day 19 15:54

First Quarter near 12 22:26; Last Quarter near 26 11:28

February – Moon and Planets

The Moon

The Moon is at its most distant apogee on February 5. It is 0.7°N of *Saturn* on February 2 and is 1.2°S of Venus later that day. On February 14, it is 1.7°N of *Aldebaran*, and five days later (February 19) is 2.4°N of *Regulus*. It is 2.5°N of *Jupiter* on February 27.

Occultations

Of the bright stars near the ecliptic that may be occulted by the Moon (*Aldebaran*, *Antares*, *Pollux*, *Regulus* and *Spica*), none are occulted in 2019. (There are occultations of *Saturn* in 2019, visible from the southern hemisphere, but not of any other planets.)

The planets

Mercury is lost in daylight, and although it reaches greatest eastern elongation on February 27, it is too low to be readily visible. *Venus* is visible in the morning sky, rapidly moving closer to the Sun and fading slightly from mag. -4.3 to -4.1 over the month. *Mars* (at mag. 0.9 to 1.2) moves from *Pisces* into *Aries* in mid-month. *Jupiter* (mag. -1.9 to -2.0) is slowly moving east in *Ophiuchus*. *Saturn*, in *Sagittarius*, is also moving very slowly eastwards at mag. 0.6. *Uranus* (mag. 5.8) is in *Aries*, just inside the border with *Pisces*. *Neptune* (mag. 7.9) remains in *Aquarius*.

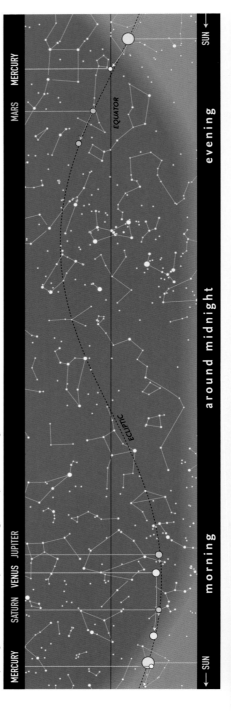

The path of the Sun and the planets along the ecliptic in February.

Calendar for February

02	07:18	Saturn 0.7°S of Moon
02	21:27	Venus 1.2°N of Moon
04	21:04	New Moon
05	07:02	Mercury 6.1°N of Moon
05	09:29	Moon at apogee (farthest of year, 406,555 km)
07	12:37	Mercury 8.4°N of Moon
10	16:19	Mars 6.1°N of Moon
12	22:26	First Quarter
14	02:29	Aldebaran 1.7°S of Moon
18	03:05	Moon 0.6°S of Praesepe (Beehive cluster)
18	14:16	Venus 1.1°N of Saturn
19	09:03	Moon at perigee (closest of year, 356,761 km)
19	13:08	Regulus 2.4°S of Moon
19	15:54	Full Moon
26	11:28	Last Quarter
27	01:25	Mercury at greatest elongation (18.1°E, mag. −0.5)
27	14:16	Jupiter 2.5°S of Moon

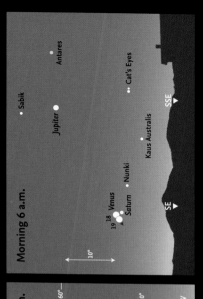

Evening 8:30 p.m.

Pleiades · Aldebaran · Moon
60°
10°
SSW · SSW

February 14 • *The Moon is very close to Aldebaran, very high in the southwest.*

Morning 6 a.m.

Algieba · Regulus · Moon
10°
WNW · W

February 19 • *The Moon is close to Regulus and Algieba when they set in the west.*

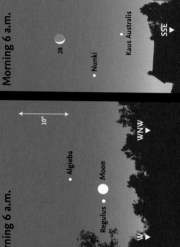

Morning 6 a.m.

Sabik · Antares · 18 Venus · 19 · Saturn · Nunki · Kaus Australis · Jupiter · Cat's Eyes
10°
SE · SSE

February 18–19 • *Venus passes north of Saturn. Jupiter is higher and farther south. The "Cat's Eyes" are λ and υ Scorpii.*

Morning 6 a.m.

Sabik · 25 · 26 · Antares · Jupiter · 27 · Cat's Eyes · Nunki · 28 · Kaus Australis
10°
SSE · S

February 26–28 • *The Moon passes Antares and Jupiter. Nunki, Kaus Australis and the Cat's Eyes are lower.*

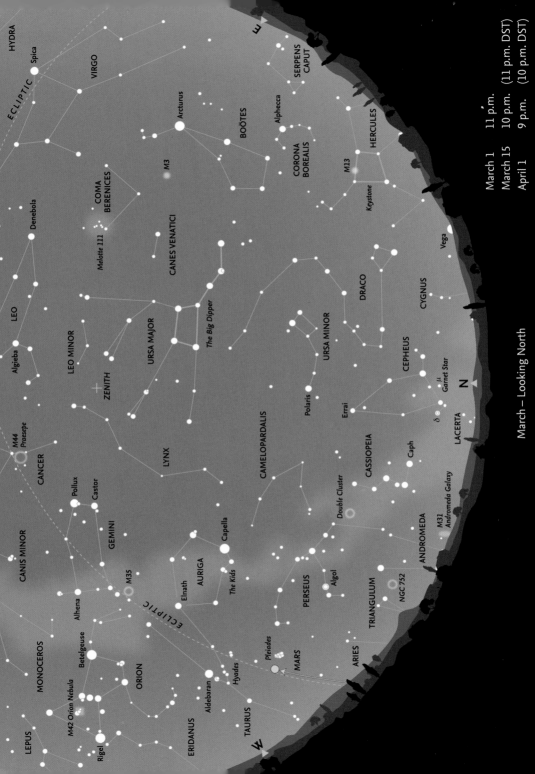

March – Looking North

March 1   11 p.m.
March 15   10 p.m.   (11 p.m. DST)
April 1   9 p.m.   (10 p.m. DST)

March – Looking North

In March, the Sun crosses the celestial equator on Wednesday, March 20, at the vernal equinox, when day and night are of almost exactly equal length, and the season of spring is considered to have begun. (The hours of daylight and darkness change most rapidly around the equinoxes, in March and September.) It is also in March that Daylight Saving Time (DST) begins in North America (on Sunday, March 10) so the charts show the appearance at 11 p.m. for March 1 and 10 p.m. DST for April 1. (In Europe, Summer Time is introduced three weeks later, on Sunday, March 31.)

Early in the month, the constellation of **Cepheus** lies almost due north, with the distinctive "W" of **Cassiopeia** to its west. Cepheus lies across the border of the Milky Way and is often described as like the gable-end of a house and is best seen later in the night or in the month. Despite the large number of stars revealed at the base of the constellation by binoculars, one star stands out because of its deep red color. This is Mu (μ) Cephei, also known as the **Garnet Star**, because of its striking color. It is a truly gigantic star, a red supergiant, and one of the largest stars known. It is about 2,400 times the diameter of the Sun, and if placed in the Solar System would extend beyond the orbit of Saturn. (Betelgeuse, in Orion, is also a red supergiant, but it is "only" about 500 times the diameter of the Sun.)

Another famous, and very important star in Cepheus is δ **Cephei**, which is the prototype for the class of variable stars known as Cepheids. These giant stars show a regular variation in their luminosity, and there is a direct relationship between the period of the changes in magnitude and the stars' actual luminosity. From a knowledge of the period of any Cepheid, its actual luminosity – known as its absolute magnitude – may be derived. A comparison of its apparent magnitude on the sky and its absolute magnitude enables the star's exact distance to be

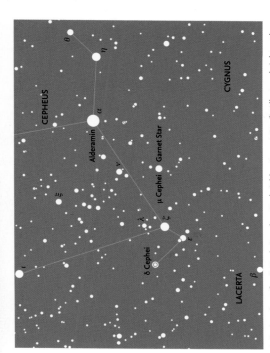

A finder chart for δ Cephei and μ Cephei (the Garnet Star). All stars brighter than magnitude 7.5 are shown.

determined. Once the distances to the first Cepheid variables had been established, examples in more distant galaxies provided information about the scale of the universe. Cepheid variables are the first "rung" in the cosmic distance ladder. Both important stars are shown on the accompanying chart.

Below Cepheus to the east (to the right), late in the night, it may be possible to catch a glimpse of **Deneb** (α Cygni), just above the horizon. Slightly farther round toward the northeast, **Vega** (α Lyrae) is marginally higher in the sky. From most of the United States, Deneb is just far enough north to be seen at some time during the night (although difficult to see in January and February because it is so low). Vega, by contrast, farther south, is completely hidden during the depths of winter.

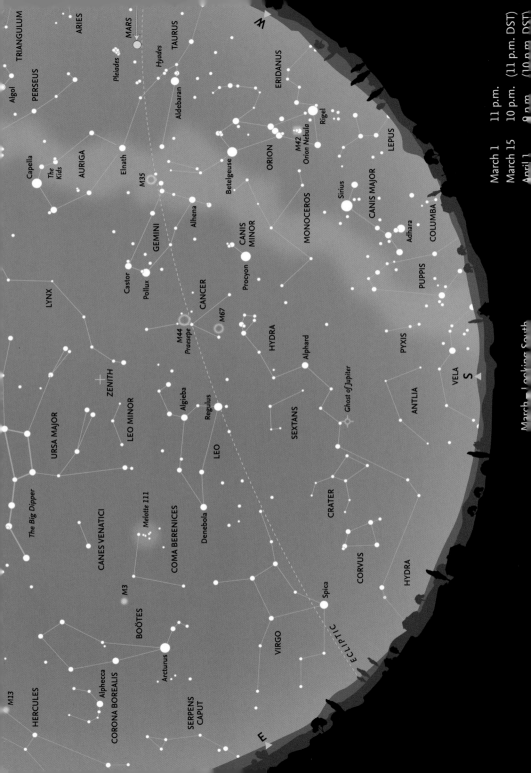

March — Looking South

March 1   11 p.m.
March 15   10 p.m.   (11 p.m. DST)
April 1   9 p.m.   (10 p.m. DST)

March – Looking South

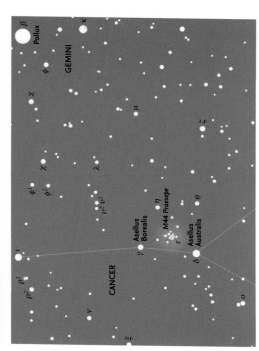

Due south at 10 p.m. at the beginning of the month, lying between the constellations of **Gemini** in the west and **Leo** in the east, and fairly high in the sky above the head of Hydra, is the faint, and rather undistinguished zodiacal constellation of **Cancer**. Rather like an upside-down letter "Y," it has three "legs" radiating from the center, where there is an open cluster, M44 or **Praesepe** ("the Manger," but also known as "the Beehive"). On a clear night this cluster, known since antiquity, is just a hazy spot to the naked eye, but appears in binoculars as a group of dozens of individual stars.

Also prominent in March is the constellation of **Leo**, with the "backward question mark" (or "Sickle") of bright stars forming the head of the mythological lion. **Regulus** (α Leonis) – the "dot" of the "question mark" or the handle of the sickle and the brightest star in Leo – lies very close to the ecliptic and is one of the few first-magnitude stars that may be occulted by the Moon. However, there are no occultations of Regulus in 2019, nor of any of the other four bright stars near the ecliptic.

The Moon's phases for March

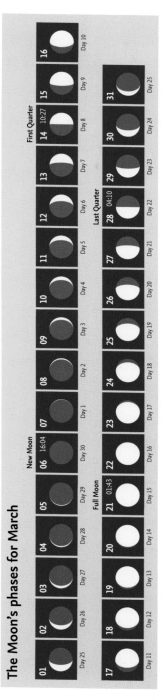

A finder chart for the open cluster M44 in Cancer. To the ancient Greeks and Romans, the two stars Asellus Borealis and Asellus Australis represented two donkeys feeding from Praesepe ("the Manger"). All stars brighter than magnitude 7.5 are shown.

March – Moon and Planets

The Moon

The Moon (in *Sagittarius*) passes close to *Venus* (just within *Capricornus*) in the morning sky on March 2. (The two bodies are closest later in the evening in Capricornus, when below the horizon.) At New, on March 6, the Moon is in *Aquarius* (as is the Sun). It is north of *Aldebaran* in *Taurus* on March 11 and *Regulus* in *Leo* on March 19. It is in *Virgo*, lying almost half-way between *Spica* and Regulus, at Full Moon on March 21. It passes close to *Jupiter* on March 27 and *Saturn* on both March 1 and March 29.

The planets

Mercury is lost in daylight (it passes inferior conjunction on March 15). *Venus* at mag. -4.0 is visible in the morning sky, crossing *Capricornus* into *Aquarius*, close to the Sun at the end of the month. *Mars* is an evening object, initially mag. 1.2 and in *Aries*, moves into *Taurus* and *Taurus* on March 11 and *Regulus* in *Leo* on March 19. It is in *Virgo*, lying fades slightly (to mag. 1.4) by the end of the month. *Jupiter* (mag. -2.0 to -2.2) is moving slowly eastwards in Ophiuchus. *Saturn* (mag. 0.6) is also moving slowly eastwards in *Sagittarius*. *Uranus* is mag. 5.9 in *Aries*, and *Neptune* (mag. 7.9) remains in *Aquarius*.

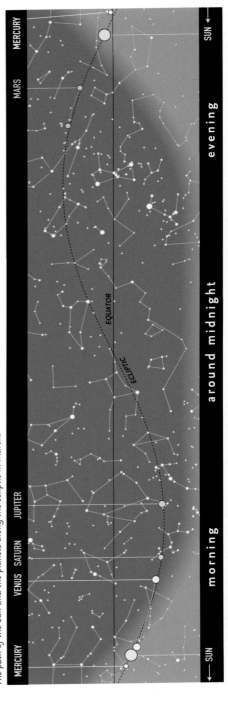

The path of the Sun and the planets along the ecliptic in March.

Calendar for March

01	18:40	Saturn 0.3°S of Moon
02	21:28	Venus 1.3°N of Moon
04	11:26	Moon at apogee (406,391 km)
06	16:04	New Moon
10		Daylight Saving Time begins (North America)
11	12:09	Mars 5.7°N of Moon
13	11:13	Aldebaran 2°S of Moon
14	10:27	First Quarter
15	01:47	Mercury at inferior conjunction
19	00:59	Regulus 2.5°S of Moon
19	19:48	Moon at perigee (359,377 km)
20	21:58	Northern spring equinox
21	01:43	Full Moon
27	03:28	Jupiter 2°S of Moon
28	04:10	Last Quarter
29	06:11	Saturn 0.1°N of Moon
31		Summer Time begins (Europe)
31	04:04	Mars 3.2°S of Pleiades

Morning 6 a.m.

March 1–3 • The Moon passes Nunki, Saturn and Venus, shortly before sunrise.

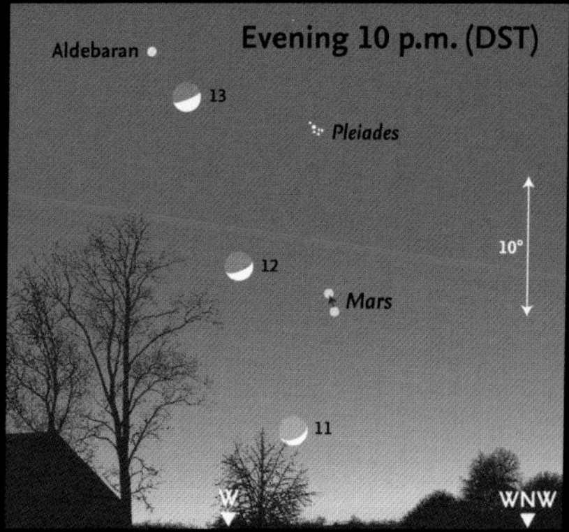

Early morning 3 a.m. (DST)

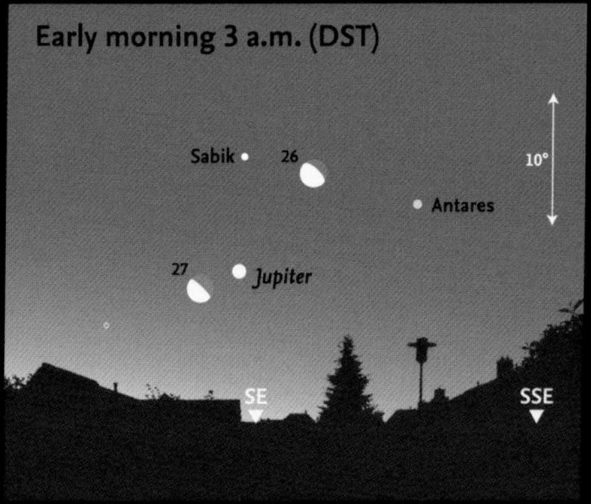

March 26–27 • Early in the morning the Moon passes Antares, Sabik and Jupiter.

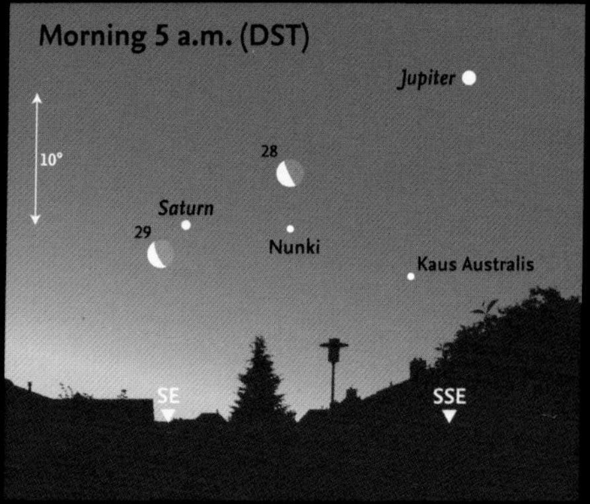

Evening 10 p.m. (DST)

March 11–13 • The Moon in the western sky, with Mars, the Pleiades and Aldebaran.

Morning 5 a.m. (DST)

March 28–29 • The Moon in the company of Nunki and Saturn. Jupiter is higher and farther south.

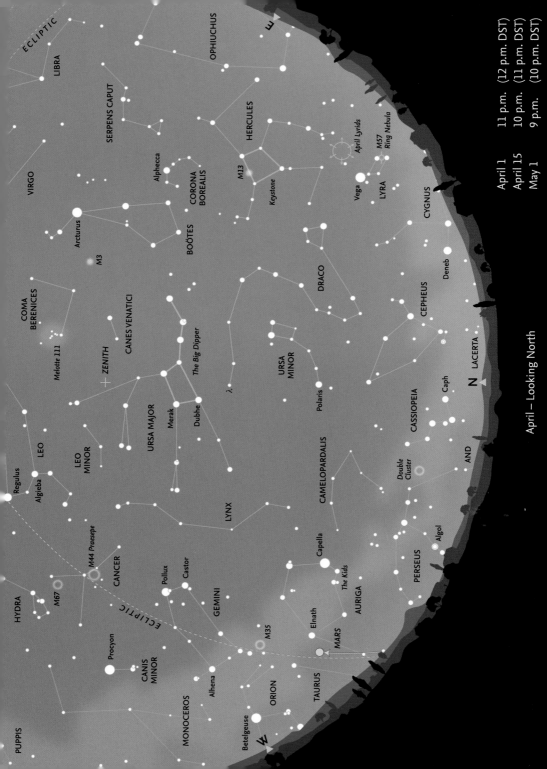

April – Looking North

April 1   11 p.m.   (12 p.m. DST)
April 15   10 p.m.   (11 p.m. DST)
May 1   9 p.m.   (10 p.m. DST)

April – Looking North

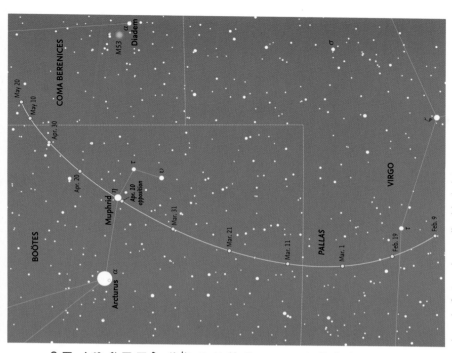

The path of minor planet Pallas (2) which is at opposition (mag. 7.9) on April 10. Background stars are shown down to magnitude 8.5.

Cygnus and the brighter regions of the Milky Way are now starting to become visible later in the night. Rising in the northeast is the small constellation of **Lyra** and the distinctive "Keystone" of **Hercules** above it. This asterism is very useful for locating the bright globular cluster M13 (see map on page 53), which lies on one side of the quadrilateral. The winding constellation of **Draco** weaves its way from the quadrilateral of stars that marks its "head," on the border with Hercules, to end at λ Draconis between **Polaris** (α Ursae Minoris) and the "Pointers," **Dubhe** and **Merak** (α and β Ursae Majoris, respectively). **Ursa Major** is "upside down" high overhead, near the zenith. The constellation of **Gemini** stands almost vertically in the west. **Auriga** is still clearly seen in the northwest, but, by the end of the month, the southern portion of **Perseus** is starting to dip below the northern horizon. The very faint constellation of **Camelopardalis** lies in the northwest between Polaris and the constellations of Auriga and Perseus.

Meteors

A moderate meteor shower, the **Lyrids**, peaks on April 21–22. Although the hourly rate is not very high (about 18 meteors per hour), the meteors are fast and some leave persistent trains. This year the maximum occurs shortly after the Full Moon, meaning conditions are not ideal for seeing the fainter meteors. The parent object is the non-periodic comet C/1861 G1 (Thatcher). Another, stronger shower, the **Eta Aquariids**, begins to be active around April 19, and comes to maximum in May.

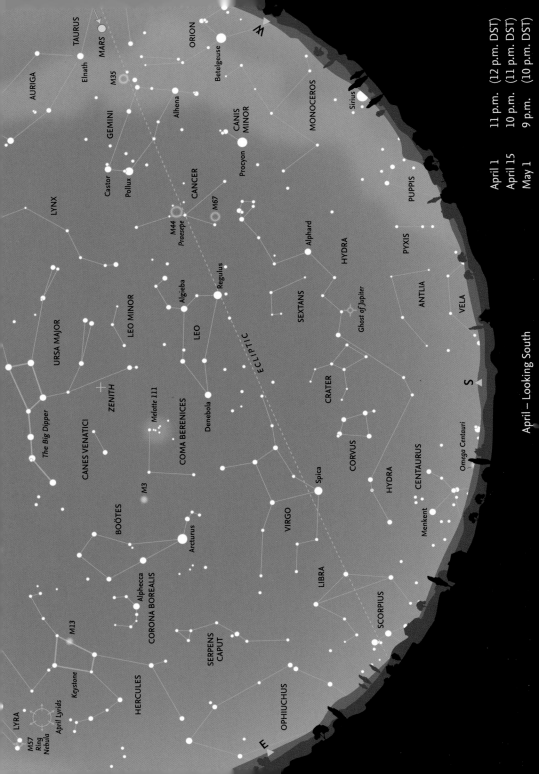

April – Looking South

April 1	11 p.m. (12 p.m. DST)
April 15	10 p.m. (11 p.m. DST)
May 1	9 p.m. (10 p.m. DST)

April – Looking South

Leo is the most prominent constellation in the southern sky in April, and vaguely looks like the creature after which it is named. *Gemini*, with *Castor* and *Pollux*, remains clearly visible in the west, and *Cancer* lies between the two constellations. To the east of Leo, the whole of *Virgo*, with *Spica* (α Virginis) its brightest star, is well clear of the horizon and Libra is coming into view. Below Leo and Virgo, the complete length of *Hydra* is visible, running beneath both constellations, with *Alphard* (α Hydrae) halfway between Regulus and the southwestern horizon. Farther east, the two small constellations of *Crater* and the rather brighter *Corvus* lie between Hydra and Virgo. Part of *Centaurus* is now above the horizon.

Boötes and *Arcturus* are prominent in the eastern sky, together with the circlet of *Corona Borealis*, framed by Boötes and the neighboring constellation of *Hercules*. Between Leo and Boötes lies the constellation of *Coma Berenices*, notable for being the location of the open cluster Melotte 111 (see page 27) and the Coma Cluster of galaxies (Abell 1656). There are about 1,000 galaxies in this cluster, which is located near the North Galactic Pole, where we are looking out of the plane of the Galaxy and are thus able to see deep into space. Only about 10 of the brightest galaxies in the Coma Cluster are visible with the largest amateur telescopes.

The distinctive constellation of Leo, with Regulus and "The Sickle" on the west. Algieba (γ Leonis), north of Regulus, appearing double, is a multiple system of four stars.

The Moon's phases for April

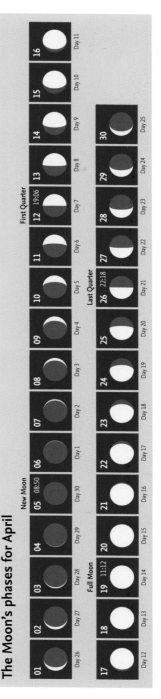

April – Moon and Planets

The Moon

At New Moon on April 5, the Moon is in *Cetus*, below the Sun, which lies in *Pisces*. On April 9 it is visible in the western sky, north of *Aldebaran* in *Taurus*. On April 15 it passes *Regulus* in *Leo*. At Full Moon, on April 19, it is in *Virgo*, not far from *Spica*. On April 23, the Moon passes close to *Jupiter* and then *Saturn* on April 25. Both planets are fairly low in the southeast.

The planets

Mercury begins in *Aquarius* and rapidly moves to greatest western elongation on April 11. *Venus* also begins the month in Aquarius, at mag. -3.8, but then moves into the dawn twilight. Both planets are too low to be readily visible. *Mars*, in the evening sky, initially mag. 1.4, fades slightly to mag. 1.6 as it moves west in *Taurus* over the month. *Jupiter*, moving slowly in *Ophiuchus*, brightens slightly from mag. -2.2 to -2.4. and *Saturn* is in *Sagittarius* at mag. 0.6–0.5. Both planets are readily visible later in the night. *Uranus* is in *Aries* at mag. 5.9 and *Neptune* in *Aquarius* at mag. 8.0. Both planets are invisible in the daytime sky. Minor planet *Pallas* (2) comes to opposition in *Boötes* on April 10 at mag. 7.9 (see page 39).

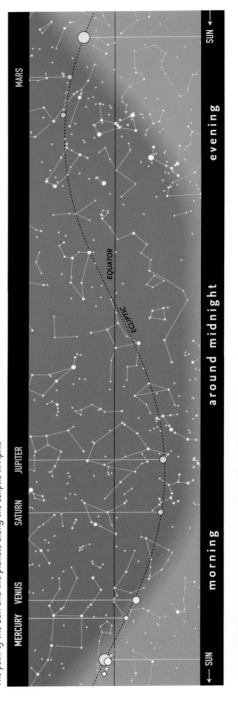

The path of the Sun and the planets along the ecliptic in April.

Calendar for April

01	00:14	Moon at apogee (405,577 km)
02	04:17	Venus 2.7°N of Moon
03	23:01	Mercury 3.6°N of Moon
05	08:50	New Moon
06	22:00 *	Mars 6.5°N of Aldebaran
09	07:40	Mars 5°N of Moon
09	16:43	Aldebaran 2.2°S of Moon
10	01:17	Pallas at opposition (mag. 7.9)
11	19:42	Mercury at greatest elongation (27.7°W, mag. 0.3)
12	19:06	First Quarter
15	01:08	Aldebaran 6.5°N of Mars
15	09:22	Regulus 2.7°S of Moon
16–25		Lyrid meteor shower
16	19:03	Mercury 4.3°S of Venus
16	22:05	Moon at perigee (364,205 km)
19–May 26		Eta Aquariid meteor shower
19	11:12	Full Moon
21–22		Lyrid shower maximum
23	11:35	Jupiter 1.6°S of Moon
25	14:27	Saturn 0.4°N of Moon
26	22:18	Last Quarter
28	18:20	Moon at apogee (404,582 km)

* These objects are close together for an extended period around this time.

Evening 10 p.m. (DST)

April 8–9 • The Moon with the Pleiades, Mars and Aldebaran. Bellatrix (γ Ori) is nearby.

Early morning 3:30 a.m. (DST)

April 15 • The Moon with Regulus and Algieba, near the western horizon.

Evening 11:30 p.m. (DST)

April 10–12 • The Moon passes Alhena and on April 12 it lines up with Castor and Pollux.

Morning 5 a.m. (DST)

April 22–26 • The Moon passes Antares, Sabik, Jupiter, Nunki and Saturn. Kaus Australis and the Cat's Eyes are below Jupiter.

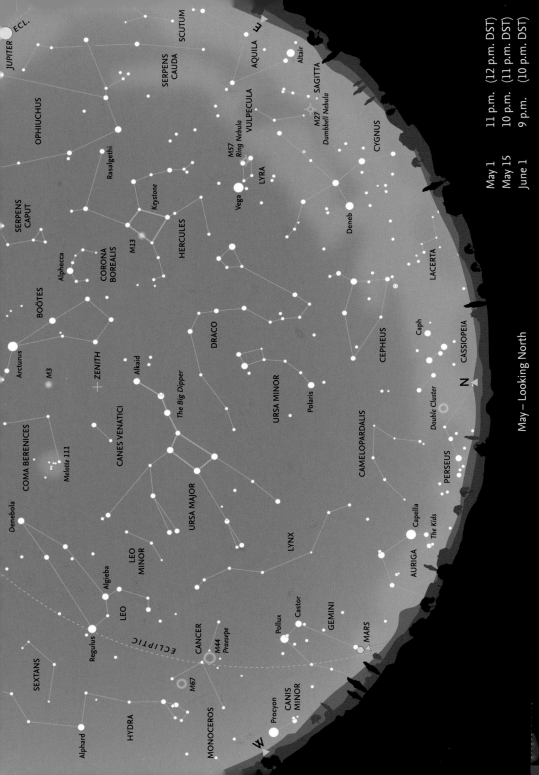

May – Looking North

May 1	11 p.m.	(12 p.m. DST)
May 15	10 p.m.	(11 p.m. DST)
June 1	9 p.m.	(10 p.m. DST)

May – Looking North

Cassiopeia is now low over the northern horizon and, to its west, the southern portions of both **Perseus** and **Auriga** have been lost below the horizon, although the **Double Cluster**, between Perseus and Cassiopeia is still clearly visible. The constellations of **Lyra, Cepheus, Ursa Minor** and the whole of **Draco** are well placed in the sky. **Gemini**, with **Castor** and **Pollux**, is sinking towards the western horizon. **Capella** (α Aurigae) and the asterism of **The Kids** are still just clear of the horizon.

In the east, two of the stars of the "Summer Triangle," **Vega** (α Lyrae) and **Deneb** (α Cygni), are clearly visible, and the third star, **Altair** in **Aquila**, is beginning to climb above the horizon. The whole of **Cygnus** is now visible. The sprawling constellation of **Hercules** is high in the east and the brightest globular cluster in the northern hemisphere, M13, is visible to the naked eye on the western side of the asterism known as the **Keystone**.

Three faint constellations may be identified before the lighter nights of summer make them difficult objects. Below Cepheus, low in the northeastern sky is the zig-zag constellation of **Lacerta**, while to the west, above Perseus and Auriga is **Camelopardalis** and, farther west, the line of faint stars forming **Lynx**.

Later in the night (and in the month) the westernmost stars of **Pegasus** begin to come into view, while the stars of **Andromeda** start to appear on the northeastern horizon. High overhead, **Alkaid** (η Ursae Majoris), the last star in the "tail" of the Great Bear, is close to the zenith, while the main body of the constellation has swung round into the western sky.

Meteors

The **Eta Aquariids** are one of the two meteor showers associated with Comet 1P/Halley (the other being the **Orionids**, in October). The Eta Aquariids are not particularly favorably placed for northern-hemisphere

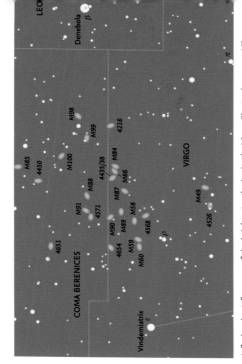

A finder chart for some of the brightest galaxies in the Virgo Cluster (see page 46). All stars brighter than magnitude 8.5 are shown.

observers, because the radiant is near the celestial equator, near the "Water Jar" in **Aquarius**, well below the horizon until late in the night (around dawn). However, meteors may still be seen in the eastern sky even when the radiant is below the horizon. There is a radiant map for the Eta Aquariids on page 16.

Their maximum in 2019, on May 6–7, occurs when the Moon is a narrow waxing crescent, just after New Moon, so conditions are particularly favorable. Maximum hourly rate is about 55 per hour and a large proportion (about 25 per cent) of the meteors leave persistent trains.

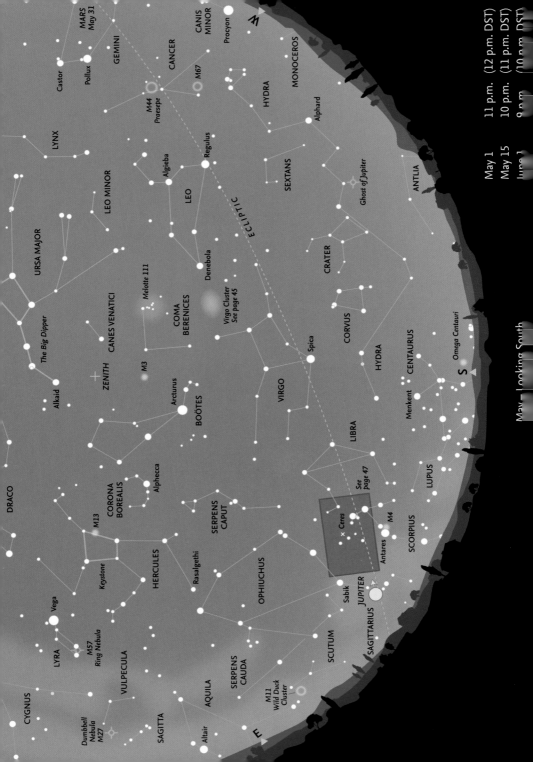

May – Looking South

May 1	11 p.m.	(12 p.m. DST)
May 15	10 p.m.	(11 p.m. DST)
June 1	9 p.m.	(10 p.m. DST)

May – Looking South

Early in the night, the constellation of *Virgo*, with *Spica* (α Virginis), lies due south, with *Leo* and both *Regulus* and *Denebola* (α and β Leonis, respectively) to its west high in the sky. The rather faint zodiacal constellation of *Libra* is now fully visible, together with most of *Scorpius* and ruddy *Antares* (α Scorpii). In the south, more of *Centaurus* may be seen, together with part of *Lupus.*

Virgo contains the nearest large cluster of galaxies, which is the center of the Local Supercluster, of which the Milky Way galaxy forms part. The Virgo Cluster contains some 2,000 galaxies, the brightest of which are visible in amateur telescopes.

Arcturus in *Boötes* is high in the south, with the distinctive circlet of *Corona Borealis* clearly visible to its east. The brightest star (α Coronae Borealis) is known as *Alphecca.* The large constellation of *Ophiuchus* (which actually crosses the ecliptic, and is thus the "13th" zodiacal constellation) is climbing into the eastern sky. Before the constellation boundaries were formally adopted by the International Astronomical Union in 1930, the southern region of Ophiuchus was regarded as

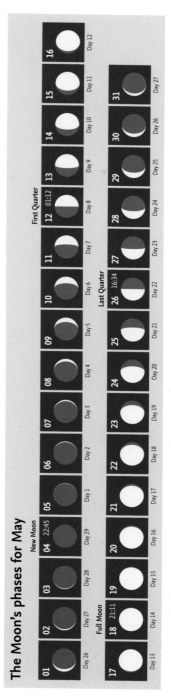

The path of minor planet Ceres (1) which comes to opposition (mag. 7.0) on May 28. Background stars are shown down to magnitude 8.5.

forming part of the constellation of Scorpius, which had been part of the zodiac since antiquity.

The Moon's phases for May

New Moon

01	02	03	04 22:45	05	06	07	08	09	10	11	12 01:12	13	14	15	16
Day 26	Day 27	Day 28	Day 29	Day 1	Day 2	Day 3	Day 4	Day 5	Day 6	Day 7	Day 8	Day 9	Day 10	Day 11	Day 12

First Quarter

17	18 21:11	19	20	21	22	23	24	25	26 16:34	27	28	29	30	31
Day 13	Day 14	Day 15	Day 16	Day 17	Day 18	Day 19	Day 20	Day 21	Day 22	Day 23	Day 24	Day 25	Day 26	Day 27

Full Moon — Last Quarter

May – Moon and Planets

The Moon

New Moon occurs on May 4, when it is actually in the constellation of **Aries**, close to the border with **Cetus**. It is visible near to **Mars** on May 7 just before the planet sets in the west. It appears in **Leo**, not far from **Regulus** on May 12, at First Quarter. Full Moon is on May 18, in the constellation of **Libra**. It is close to **Jupiter** on May 20 and passes **Saturn** on May 22.

The planets

Mercury is invisible, close to the Sun, and is at superior conjunction on May 21. **Venus** is close to the horizon at dawn at the beginning of the month, but rapidly becomes invisible in daylight. **Mars** (mag. 1.6–1.7) may be glimpsed in **Taurus** in the evening twilight early in the month, but by the end of the month becomes too low as it moves into **Gemini**. **Jupiter** (mag. -2.5), and still in **Ophiuchus**, rises about 23:00 UT. **Saturn** (mag. 0.5), in **Sagittarius**, rises slightly later at about 01:00 and brightens slightly to mag. 0.3 over the month. Both planets are lost in twilight around 04:00 UT. **Uranus** is in **Aries** at mag. 5.9 and **Neptune** in **Aquarius** at mag. 7.9. The minor planet **Ceres** is at opposition (mag. 7.0) on May 28 in **Ophiuchus**, just inside the border with **Scorpius**.

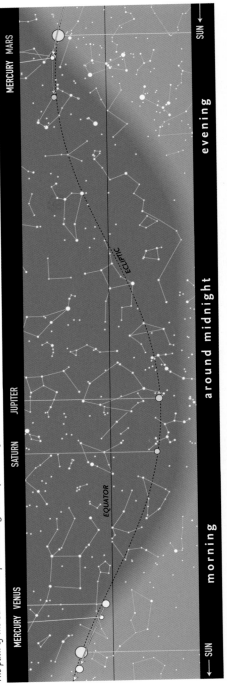

The path of the Sun and the planets along the ecliptic in May.

Calendar for May

02	11:39	Venus 3.6°N of Moon
03	06:25	Mercury 2.9°N of Moon
04	22:45	New Moon
06–07		Eta Aquariid shower maximum
06	22:20	Aldebaran 2.3°S of Moon
07	23:35	Mars 3.2°N of Moon
10	03:56	Pollux 6.3°N of Moon
12	01:12	First Quarter
12	14:44	Regulus 3°S of Moon
13	21:53	Moon at perigee (369,009 km)
16	06:37	Spica 7.7°S of Moon
18	21:11	Full Moon
19	17:05	Antares 7.9°S of Moon
20	16:54	Jupiter 1.7°S of Moon
21	13:07	Mercury at superior conjunction
22	22:14	Saturn 0.5°N of Moon
26	13:27	Moon at apogee (404,138 km)
26	16:34	Last Quarter
28	22:36	Ceres at opposition (mag. 7.0)

Evening 8:30 p.m. (DST)

May 6 • The Moon with Aldebaran, Mars and Elnath, shortly after sunset.

Evening 10 p.m. (DST)

May 7–9 • The Moon passes Mars, Elnath, Alhena and Pollux. Castor and Procyon are also nearby.

After midnight 1 a.m. (DST)

May 12 • The Moon passes Regulus and Algieba, near the western horizon.

After midnight 2 a.m. (DST)

May 19–23 • The Moon passes Antares, Sabik, Jupiter, Nunki and Saturn, in the southern sky.

MAY

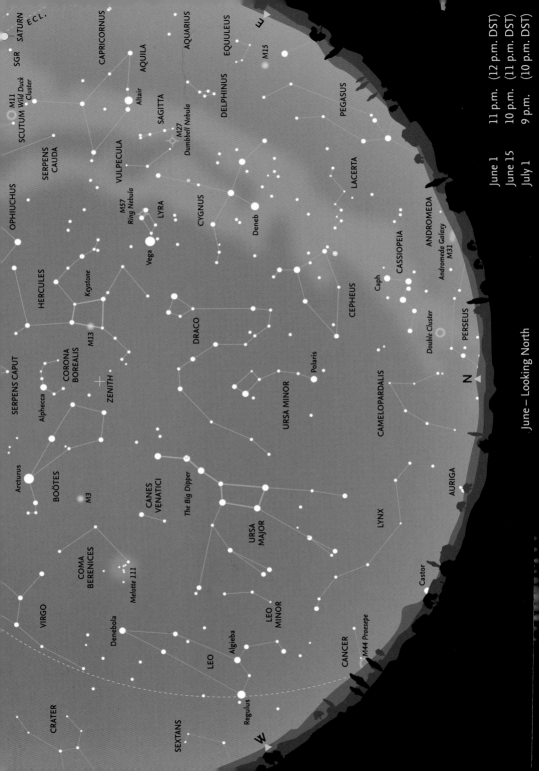

June – Looking North

June 1	11 p.m.	(12 p.m. DST)
June 15	10 p.m.	(11 p.m. DST)
July 1	9 p.m.	(10 p.m. DST)

June – Looking North

With the approach to the summer solstice (June 21), twilight tends to persist in the northern United States and Canada, with four hours of darkness in Toronto in June; two hours in Seattle; and none in Vancouver. As a result, most of the fainter stars and constellations are invisible. Even brighter stars, such as the seven stars making up the well-known asterism of the **Big Dipper** in **Ursa Major**, may be difficult to detect except around local midnight (1 a.m. DST). In the south of the United States, at New Orleans, Houston or Los Angeles, there are six or even seven hours of true darkness. Northern observers may have the compensation of seeing noctilucent clouds (NLC), electric-blue clouds visible during summer nights in the direction of the North Pole, for about a month or six weeks on either side of the solstice.

Two faint constellations, **Camelopardalis** and **Lynx**, may be glimpsed low on the northern and western horizons. **Leo** and **Regulus** (α Leonis) are heading downwards in the western sky, but **Cassiopeia**, farther east, is low, but clearly visible. The stars of the "Summer Triangle" (**Vega, Deneb** and **Altair**) in the constellations of **Lyra, Cygnus** and **Aquila**, respectively, are now readily seen in the east, and the small, distinctive constellation of **Delphinus** is well clear of the horizon.

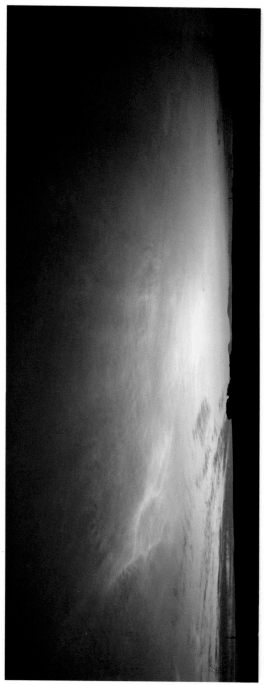

Noctilucent clouds, photographed on the night of July 5–6, 2016, from Portmahomack, Ross-shire, Scotland, by Denis Buczynski.

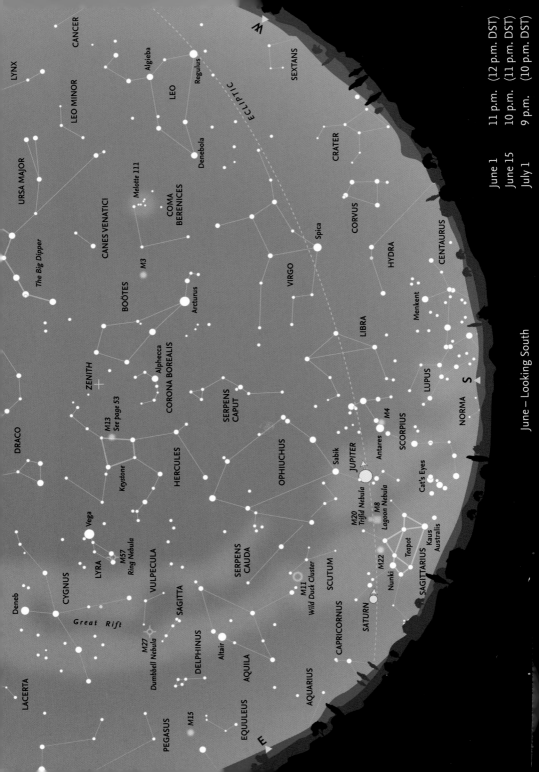

June – Looking South

June 1	11 p.m.	(12 p.m. DST)
June 15	10 p.m.	(11 p.m. DST)
July 1	9 p.m.	(10 p.m. DST)

LYNX

CANCER

LEO MINOR

URSA MAJOR

LEO

Algieba

Regulus

SEXTANS

Denebola

CRATER

The Big Dipper

Melotte 111

COMA BERENICES

CORVUS

CANES VENATICI

M3

HYDRA

CENTAURUS

Spica

BOÖTES

VIRGO

Arcturus

Menkent

DRACO

ZENITH

Alphecca

CORONA BOREALIS

LIBRA

SERPENS CAPUT

LUPUS

M13
See page 53

S

HERCULES

NORMA

Keystone

SCORPIUS

M4

OPHIUCHUS

Antares

Sabik

JUPITER

Vega

Cat's Eyes

LYRA

M57
Ring Nebula

M20
Trifid Nebula

M8
Lagoon Nebula

CYGNUS

VULPECULA

Kaus
Australis

Deneb

SERPENS CAUDA

Teapot

SAGITTA

SCUTUM

M22

Great Rift

M11
Wild Duck Cluster

Nunki

SAGITTARIUS

M27
Dumbbell Nebula

SATURN

DELPHINUS

CAPRICORNUS

LACERTA

Altair

AQUILA

PEGASUS

EQUULEUS

AQUARIUS

M15

E

ECLIPTIC

June – Looking South

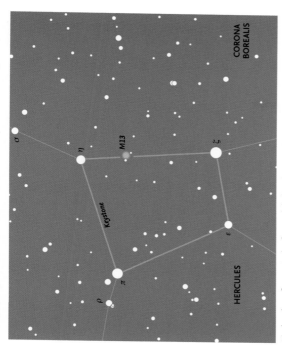

The rather undistinguished constellation of *Libra* now lies almost due south. The red supergiant star *Antares* – the name means the "Rival of Mars" – in *Scorpius* is visible slightly to the east of the meridian, and even the "tail" or "sting" is visible with clear skies. Larger areas of *Centaurus* and *Lupus* are visible in the south, with *Sagittarius* to the east. Higher in the sky is the large constellation of *Ophiuchus* (the "Serpent Bearer"), lying between the two halves of *Serpens: Serpens Caput* ("Head of the Serpent") to the west and *Serpens Cauda* ("Tail of the Serpent") to the east. (Serpens is the only constellation to be divided into two distinct parts.) The ecliptic runs across Ophiuchus, and the Sun spends far more time in the constellation than it does in the "classical" zodiacal constellation of Scorpius, a small area of which lies between Libra and Ophiuchus.

Higher in the southern sky, the three constellations of *Boötes*, *Corona Borealis* and *Hercules* are now better placed for observation than at any other time of the year. This is an ideal time to observe the fine globular cluster of M13 in Hercules.

The Moon's phases for June

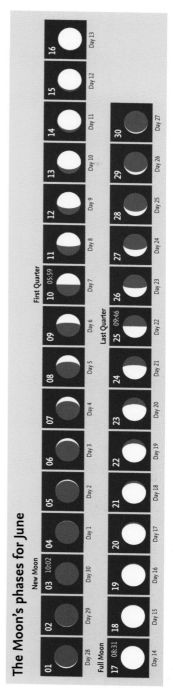

Finder chart for M13, the finest globular cluster in the northern sky. All stars down to magnitude 7.5 are shown.

June – Moon and Planets

The Moon

New Moon is on June 3 when it is close to **Aldebaran** in **Taurus**, but the star is invisible in morning twilight. Later in the month the Moon is again near Aldebaran on June 30, but the star is now in daylight. It passes close to **Mars** on June 5, and both bodies may be glimpsed later in **Gemini** just before they set in the west. It passes **Spica** in **Virgo** in daylight on June 12 and again both bodies may be seen later in the day as they begin to set in the southwest. It is in **Ophiuchus**, above **Antares** (in **Scorpius**) on June 16 and passes close to **Jupiter** later that day. Full Moon is on June 17 (in **Ophiuchus**, near Jupiter). It passes very close to **Saturn** (in **Sagittarius**) on June 19, but the event occurs in the southeast when they are below the horizon, although both bodies may be seen later in the early morning twilight.

The planets

Mercury is initially in the daytime sky, but rapidly moves to greatest eastern elongation on June 23, when it is actually close to **Mars**, but too low to be visible after sunset. It may be possible to glimpse **Venus** (at mag. -3.9) early in the month, very low in morning twilight, but it soon moves into daylight. **Mars** is in **Gemini**, but invisible in daylight, although moving into the western evening sky. **Jupiter** is in **Ophiuchus** at mag. -2.6 and comes to opposition on June 10, **Saturn** (mag. 0.3–0.1) is in **Sagittarius**. **Uranus** remains in Aries at mag. 5.9 and **Neptune** in **Aquarius** at mag. 7.9.

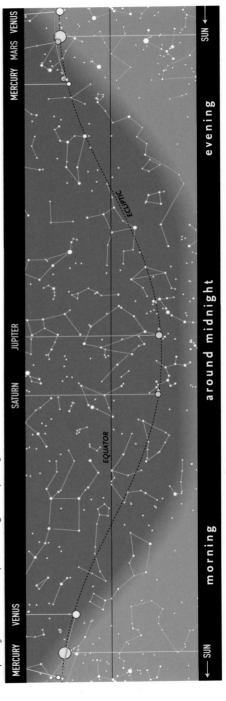

The path of the Sun and the planets along the ecliptic in June.

MERCURY   VENUS   SATURN   JUPITER   MERCURY   MARS   VENUS

SUN ◄—   morning   around midnight   evening   — ► SUN

ECLIPTIC

EQUATOR

Calendar for June

01	18:14	Venus 3.2°N of Moon
03	06:12	Aldebaran 2.3°S of Moon
03	10:02	New Moon
04	15:41	Mercury 3.7°N of Moon
05	15:05	Mars 1.6°N of Moon
06	10:07	Pollux 6.2°N of Moon
07	23:15	Moon at perigee (368,504 km)
08	20:01	Regulus 3.2°S of Moon
10	05:59	First Quarter
10	15:28	Jupiter at opposition (mag. -2.6)
12	12:49	Spica 7.8°S of Moon
16	01:02	Antares 8°S of Moon
16	18:50	Jupiter 2°S of Moon
17	08:31	Full Moon
17	21:00 *	Venus 4.8°N of Aldebaran
18	15:00 *	Mars 0.2°S of Mercury
19	03:46	Saturn 0.4°N of Moon
21	15:54	Summer solstice
23	07:00 *	Mars 5.6°S of Pollux
23	07:50	Moon at apogee (404,548 km)
23	23:16	Mercury at greatest elongation (25.2°E, mag. 0.4)
25	09:46	Last Quarter
30	15:34	Aldebaran 2.3°S of Moon

* These objects are close together for an extended period around this time.

Evening 9:30 p.m. (DST)

June 8 • The Moon with Regulus and Algieba. Denebola is a little higher.

Evening 9 p.m. (DST)

June 4–6 • The narrow crescent Moon near Mercury, Mars and Pollux, shortly after sunset in the western sky. Procyon is nearby.

Evening 9:15 p.m. (DST)

June 18 • Mercury and Mars, close together after sunset. Castor and Pollux are nearby.

After midnight 2 a.m. (DST)

June 16–19 • The Moon passes Antares, Sabik, Jupiter, Nunki and Saturn. Kaus Australis is closer to the horizon.

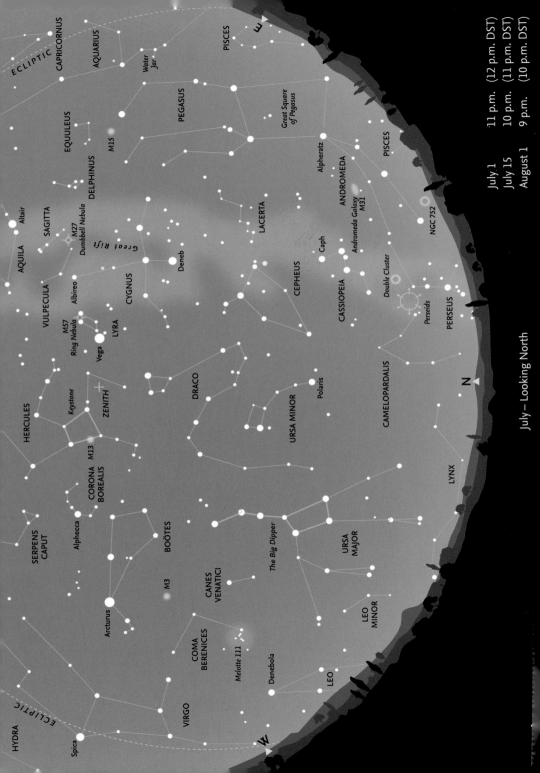

July – Looking North

July 1   11 p.m. (12 p.m. DST)
July 15   10 p.m. (11 p.m. DST)
August 1   9 p.m. (10 p.m. DST)

July – Looking North

As in June, light nights and the chance of observing noctilucent clouds persist throughout July, but later in the month (and particularly after midnight) some of the major constellations begin to be more easily seen. **Capella**, the brightest star in **Auriga** together with the rest of the constellation, is hidden below the northern horizon. **Cassiopeia** is clearly visible in the northeast and **Perseus**, to its south, is beginning to climb clear of the horizon. The band of the Milky Way, from Perseus through Cassiopeia toward **Cygnus**, stretches up into the northeastern sky. If the sky is dark and clear, you may be able to make out the small, faint constellation of **Lacerta**, lying across the Milky Way between Cassiopeia and Cygnus. In the east, the stars of **Pegasus** are now well clear of the horizon, with the main line of stars forming **Andromeda** roughly parallel to the horizon low in the northeast. **Alpheratz** (α Andromedae) is actually the star at the northeastern corner of the **Great Square of Pegasus**. **Cepheus** and **Ursa Major** are on opposite sides of **Polaris** and **Ursa Minor**, in the east and west, respectively. The head of **Draco** is very close to the zenith (which is in **Hercules**) so the whole of this winding constellation is readily seen.

Meteors

July brings increasing meteor activity, mainly because there are several minor radiants active in the constellations of **Capricornus** and **Aquarius**. Because of their location, however, observing conditions are not particularly favourable for northern-hemisphere observers, although the first shower, the **Alpha Capricornids**, active from July 11 to August 10 (peaking July 26 to August 1), does often produce very bright fireballs. The maximum rate, however, is only about 5 per hour. The parent body is Comet 169P/NEAT. The most prominent shower is probably that of the **Delta Aquariids**, which are active from around July 21 to August 23, with a peak on July 29–30, although even then the hourly rate is unlikely to reach 20 meteors per hour. In this case, the parent body is possibly Comet 96P/Machholz. This year both shower maxima occur when the Moon is a waning crescent, so observing conditions are reasonably favourable. A chart showing the **Delta Aquariid** radiant is shown on page 16. The **Perseids** begin on July 13 and peak on August 11–12.

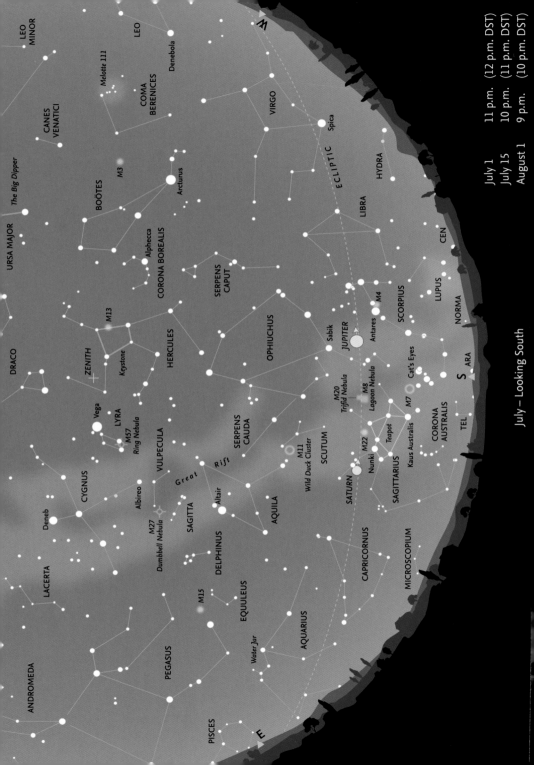

July – Looking South

July 1   11 p.m.   (12 p.m. DST)
July 15   10 p.m.   (11 p.m. DST)
August 1   9 p.m.   (10 p.m. DST)

July – Looking South

This is the best time of year to see **Scorpius**, with deep red **Antares** (α Scorpii), glowing just above the southern horizon. At around midnight, **Sagittarius**, with the distinctive asterism of the "Teapot," and the dense star clouds of the center of the Milky Way, are visible in the south, together with the small constellation of **Corona Australis**. The **Great Rift** – actually dust clouds that hide the more distant stars – runs down the Milky Way from Cygnus toward Sagittarius. Towards its northern end is the small constellation of **Sagitta** and the planetary nebula **M27** (the Dumbbell Nebula). The sprawling constellation of **Ophiuchus** lies close to the meridian for a large part of the month, separating the two halves of the constellation of **Serpens**. The western half is called **Serpens Caput** (Head of the Serpent) and the eastern part **Serpens Cauda** (Tail of the Serpent). In the east, the bright **Summer Triangle**, consisting of **Vega** in Lyra, **Deneb** in Cygnus and **Altair** in Aquila, begins to dominate the southern sky, as it will throughout August and into September. The small constellation of Lyra, with Vega and a distinctive quadrilateral of stars to its east and south, lies not far from the zenith.

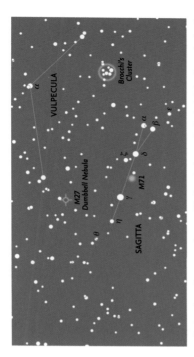

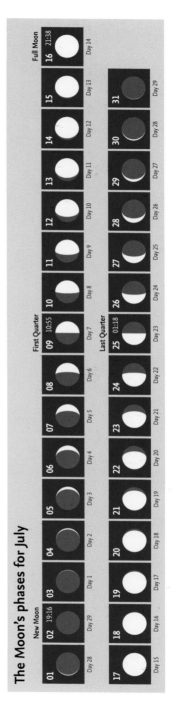

A finder chart for M27, the Dumbbell Nebula, a relatively bright (magnitude 8) planetary nebula – a shell of material ejected in the late stages of a star's lifetime – in the constellation of Vulpecula. All stars brighter than magnitude 7.5 are shown.

The Moon's phases for July

								First Quarter		Full Moon
New Moon										
01	02 19:16	03	04	05	06	07	08	09 10:55		16 21:38
Day 28	Day 29	Day 1	Day 2	Day 3	Day 4	Day 5	Day 6	Day 7		Day 14

10	11	12	13	14	15	16 21:38
Day 8	Day 9	Day 10	Day 11	Day 12	Day 13	Day 14

Last Quarter										
17	18	19	20	21	22	23	24	25 01:18		31
Day 15	Day 16	Day 17	Day 18	Day 19	Day 20	Day 21	Day 22	Day 23		Day 29

26	27	28	29	30	31
Day 24	Day 25	Day 26	Day 27	Day 28	Day 29

July – Moon and Planets

The Moon

New Moon occurs on July 2, when the Moon (and Sun) are in *Gemini*. That day there is a total solar eclipse, visible from the South Pacific and the southern tip of South America. On July 4, the Moon is very close to *Mars* in *Cancer*, but both bodies are lost in daylight. It passes *Regulus* in *Leo* on July 6, visible as the constellation sets in the west. On July 9, the Moon is close to *Spica* in *Virgo*. The objects are visible the night before as the constellation sets in the west. On July 13 the Moon is close to *Antares* in *Scorpius* and then passes *Jupiter* in *Ophiuchus*. On July 16 it lies very close to *Saturn* in *Sagittarius*, but Full Moon occurs later in the day, so Saturn will be difficult to see. That day there is a partial lunar eclipse, visible from Africa and the Middle East. On July 28, the Moon is a waning crescent close to *Aldebaran*, in *Taurus*, visible only later as the objects rise in the east.

The planets

Mercury remains invisible in daylight and passes inferior conjunction (between Earth and the Sun) on July 21. Venus is also too close to the Sun to be detected. *Mars* remains close to the Sun in *Cancer*. *Jupiter* (mag. -2.6 to -2.4) is still retrograding slowly in *Ophiuchus* and *Saturn* (mag. 0.1) is doing the same in *Sagittarius*, reaching opposition on July 9. *Uranus* is still in *Aries* at mag. 5.9 and *Neptune* in *Aquarius* at mag. 7.9.

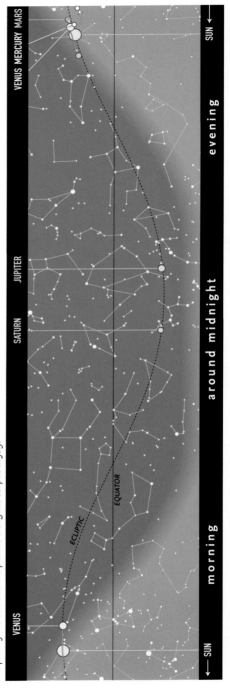

The path of the Sun and the planets along the ecliptic in July.

Calendar for July

01	21:45	Venus 1.6°N of Moon
02	19:16	New Moon
02	19:23	Total solar eclipse
		(S. Pacific, southern S. America)
03	18:24	Pollux 6.1°N of Moon
04	05:39	Mars 0.1°S of Moon
04	08:34	Mercury 3.2°S of Moon
04	22:11	Earth at aphelion
		(152,104,213 km = 1.01675 AU)
05	05:00	Moon at perigee (363,726 km)
06	02:42	Regulus 3.2°S of Moon
09	10:55	First Quarter
09	17:07	Saturn at opposition (mag. 0.1)
09	18:10	Spica 7.9°S of Moon
11–Aug.10		Alpha Capricornid meteor shower
13	07:19	Antares 8°S of Moon
13	19:43	Jupiter 2.3°S of Moon
13–Aug.26		Perseid meteor shower
16	07:15	Saturn 0.2°N of Moon
16	21:38	Full Moon
16	21:31	Partial lunar eclipse (Africa)
20	23:59	Moon at apogee (405,481 km)
21	12:34	Mercury at inferior conjunction
21–Aug.23		Delta Aquariid meteor shower
23	16:00 *	Venus 6.1°N of Pollux
25	01:18	Last Quarter
26–Aug.01		Alpha Capricornid shower maximum
28	01:16	Aldebaran 2.3°S of Moon
29–30		Delta Aquariid shower maximum
31	02:18	Mercury 4.5°S of Moon
31	04:29	Pollux 6.1°N of Moon
31	20:36	Venus 0.6°S of Moon

* These objects are close together for an extended period around this time.

Evening 10 p.m. (DST)

July 5–6 • The crescent Moon passes between Regulus and Algieba.

Evening 11:30 p.m. (DST)

July 12–16 • The waxing gibbous Moon passes Jupiter, Sabik and Antares on July 12 and 13. Two days later it is near Nunki and Saturn.

Morning 4 a.m. (DST)

July 27–28 • The Moon with Aldebaran and the Pleiades. Elnath and Capella are nearby.

Morning 5:30 a.m. (DST)

July 30 • The narrow crescent Moon is between Alhena, Castor and Pollux shortly before sunrise.

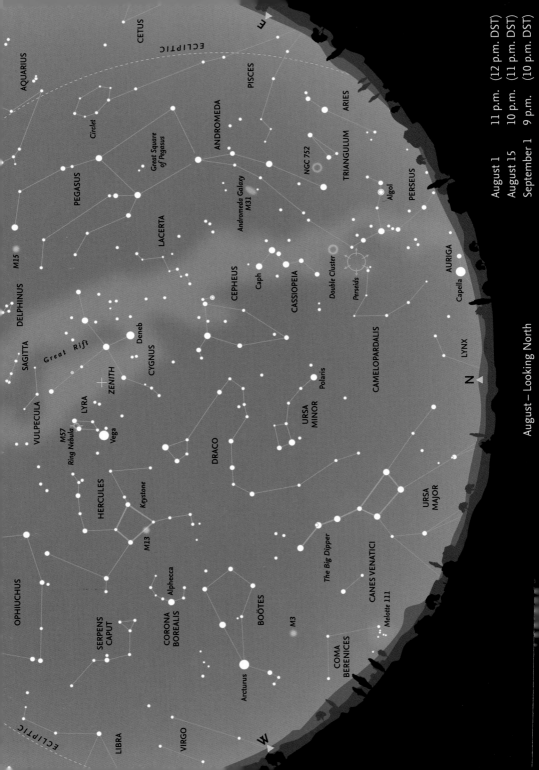

August – Looking North

August 1   11 p.m.   (12 p.m. DST)
August 15   10 p.m.   (11 p.m. DST)
September 1   9 p.m.   (10 p.m. DST)

August – Looking North

Ursa Major is now the "right way up" in the northwest, although some of the fainter stars in the south of the constellation are difficult to see. Beyond it, **Boötes** stands almost vertically in the west, but pale orange *Arcturus* is sinking toward the horizon. Higher in the sky, both **Corona Borealis** and **Hercules** are clearly visible.

In the northeast, *Capella* becomes visible later in the night, but most of *Auriga* still remains below the horizon. Higher in the sky, **Perseus** is gradually coming into full view and, later in the night and later in the month, the beautiful *Pleiades* cluster rises above the northeastern horizon. Between Perseus and *Polaris* lies the faint and unremarkable constellation of *Camelopardalis*.

Higher still, both **Cassiopeia** and **Cepheus** are well placed for observation, despite the fact that Cassiopeia is completely immersed in the band of the Milky Way, as is the "base" of Cepheus. **Pegasus** and **Andromeda** are now well above the eastern horizon and, below them, the constellation of *Pisces* is climbing into view. Two of the stars in the *Summer Triangle, Deneb* and *Vega*, are close to the zenith high overhead.

Meteors

August is the month when one of the best meteor showers of the year occurs: the **Perseids**. This is a long shower, generally beginning about July 13 and continuing until about August 26, with a maximum in 2019 on August 11–12, when the rate may reach as high as 100 meteors per hour (and on rare occasions, even higher). In 2019, maximum is just before Full Moon, so conditions are particularly unfavorable. The Perseids are debris from Comet 109P/Swift-Tuttle (the Great Comet of 1862). Perseid meteors are fast and many of the brighter ones leave persistent trains. Some bright fireballs also occur during the shower.

A brilliant Perseid fireball, streaking alongside the Great Rift in the Milky Way, photographed in 2012 by Jens Hackmann from near Weikersheim in Germany. Four additional, fainter Perseids are also visible in the image.

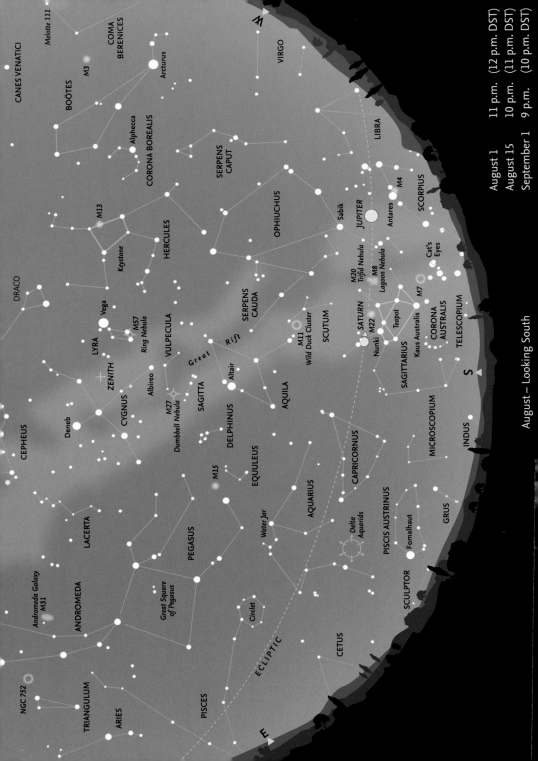

August – Looking South

August 1	11 p.m.	(12 p.m. DST)
August 15	10 p.m.	(11 p.m. DST)
September 1	9 p.m.	(10 p.m. DST)

August – Looking South

The whole stretch of the summer Milky Way stretches across the sky in the south, from **Cygnus**, high in the sky near the zenith, past **Aquila**, with bright **Altair** (α Aquilae), to part of the constellation of **Sagittarius** close to the horizon, where the pattern of stars known as the "Teapot" is visible. This area contains many nebulae and both open and globular clusters. Between **Albireo** (β Cygni) and Altair lie the two small constellations of **Vulpecula** and **Sagitta**, with the latter easier to distinguish (because of its shape) from the clouds of the Milky Way. Between Sagitta and **Pegasus** to the east lie the highly distinctive five stars that form the tiny constellation of **Delphinus** (again, one of the few constellations that actually bear some resemblance to the creatures after which they are named). Below Aquila, mainly in the star clouds of the Milky Way, lies **Scutum**, most famous for the bright open cluster, **M11** or the "Wild Duck Cluster," readily visible in binoculars. To the southeast of Aquila lie the two zodiacal constellations of **Capricornus** and **Aquarius** and, farther south, the constellation of **Piscis Austrinus** with the brilliant star **Fomalhaut**.

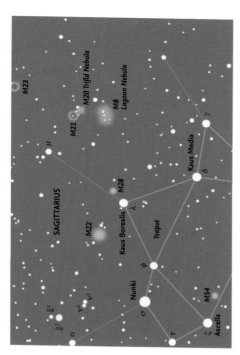

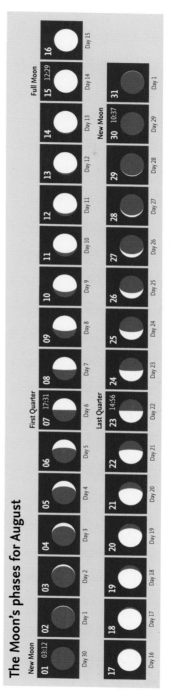

A finder chart for the gaseous nebulae M8 (the Lagoon Nebula) and M20 (the Trifid Nebula) and the globular cluster M22, all in Sagittarius. Clusters M21 & M23 (open) and M28 (globular) are faint. The chart shows all stars brighter than magnitude 7.5.

The Moon's phases for August

AUGUST 65

August – Moon and Planets

The Moon

New Moon occurs on August 1 in **Cancer**. A few hours later, it passes **Mars**, just inside **Leo** and the next day it is close to **Regulus**, but these events occur in daylight. On August 6, just before First Quarter, it passes **Spica** in **Virgo**, but the constellation is visible only as it sets in the west. On August 9, the Moon is close to both **Antares** in **Scorpius** and **Jupiter** (in **Ophiuchus**), visible just before they set. On August 12, the Moon is close to **Saturn** in **Sagittarius**. Full Moon is on August 15, on the border of **Capricornus** and **Aquarius**. On August 30, at New Moon, the Sun, **Mercury**, **Venus** and **Mars** are all clustered together in Leo.

The planets

Mercury is at greatest eastern elongation on August 9, but rapidly becomes close to the Sun. **Venus** is invisible near the Sun in daylight. It is at superior conjunction on August 14. **Mars** (in **Leo**) at mag. 1.8 is also too close to the Sun to be readily visible, but might be glimpsed early in the month as it sets in the west. **Jupiter** (mag. -2.4 to -2.2) is in **Ophiuchus**, reaches a stationary point on August 8 and then begins normal eastwards motion. **Saturn** (mag. 0.1 to 0.3) is still retrograding slowly in **Sagittarius**. **Uranus** (mag. 5.9) and **Neptune** (mag. 7.9) remain in **Aries** and **Aquarius**, respectively.

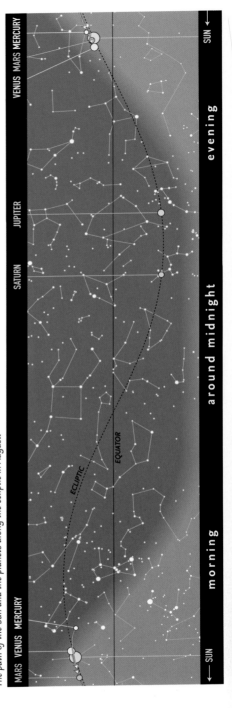

The path of the Sun and the planets along the ecliptic in August.

Calendar for August

01	03:12	New Moon
01	19:55	Mars 1.7°S of Moon
02	07:11	Moon at perigee (359,398 km)
02	11:41	Regulus 3.2°S of Moon
06	00:35	Spica 7.8°S of Moon
07	17:31	First Quarter
09	12:50	Antares 7.9°S of Moon
09	22:53	Jupiter 2.5°S of Moon
09	23:08	Mercury at greatest elongation (19.0°W, mag. -0.0)
11–12		Perseid shower maximum
12	09:53	Saturn 0.1°N of Moon (Occultation from NZ and E. Australia)
14	06:07	Venus at superior conjunction
15	12:29	Full Moon
17	10:49	Moon aof Regulus
21	04:00 *	Venus 1.0°N of Regulus
23	14:56	Last Quarter
24	09:54	Aldebaran 2.4°S of Moon
27	14:56	Pollux 6.1°N of Moon
28		Alpha Aurigid shower maximum
29	22:19	Regulus 3.2°S of Moon
30	01:07	Mercury 1.9°S of Moon
30	10:22	Mars 3.1°S of Moon
30	10:37	New Moon
30	15:53	Moon at perigee (357,176 km)
30	16:18	Venus 2.9°S of Moon

* These objects are close together for an extended period around this time.

Evening 10 p.m. (DST)

Evening 10:30 p.m. (DST)

August 5–6 • The Moon is between Spica and ζ Vir. Porrima (γ Vir) is nearby.

August 8–9 • The Moon with Acrab (β Sco), Antares Jupiter and Sabik.

Evening 10:30 p.m. (DST)

Early morning 3 a.m. (DST)

August 10–12 • The Moon moves away from Jupiter to pass Nunki and Saturn, almost due south.

August 24 • The Moon is close to Aldebaran with the Pleiades about 12 degrees higher.

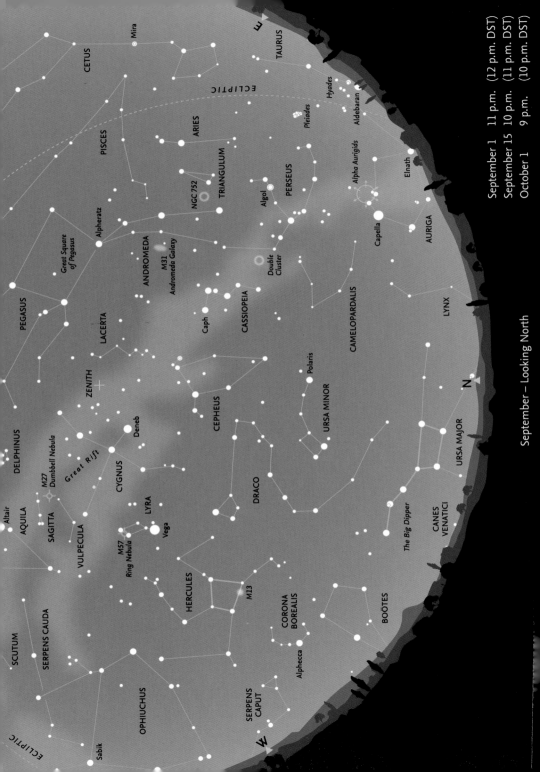

September 1   11 p.m.   (12 p.m. DST)
September 15 10 p.m.   (11 p.m. DST)
October 1   9 p.m.   (10 p.m. DST)

September – Looking North

The twin open clusters, known as the Double Cluster, in Perseus (more formally called h and χ Persei) are close to the main portion of the Milky Way.

September – Looking North

Ursa Major is now low in the north and to the northwest *Arcturus* and much of *Boötes* sink below the horizon later in the night and later in the month. In the northeast *Auriga* is beginning to climb higher in the sky. Later in the month, *Taurus*, with orange *Aldebaran* (α Tauri), and even *Gemini*, with *Castor* and *Pollux*, become visible in the east and northeast. Due east, *Andromeda* is now clearly visible, with the small constellations of *Triangulum* and *Aries* (the latter a zodiacal constellation) directly below it. Practically the whole of the Milky Way is visible, arching across the sky, both in the north and in the south. It is not particularly clear in Auriga, or even *Perseus*, but in *Cassiopeia* and on toward *Cygnus* the clouds of stars become easier to see. The *Double Cluster* in Perseus is well placed for observation. *Cepheus* is "upside-down" near the zenith, and the head of *Draco* and *Hercules* beyond it are well placed for observation.

Meteors

After the major Perseid shower in August, there is very little shower activity in September. One minor, but very extended, shower, known as the **Alpha Aurigids**, actually has two peaks of activity. The first was on August 28, but the primary peak occurs on September 15. At either of the maxima, however, the hourly rate hardly reaches 10 meteors per hour, although the meteors are bright and relatively easy to photograph. Activity from this shower also extends into October. The **Southern Taurid** shower begins this month and, although rates are low, often produces very bright fireballs. As a slight compensation for the lack of activity, however, in September the number of sporadic meteors reaches its highest rate than at any other time during the year.

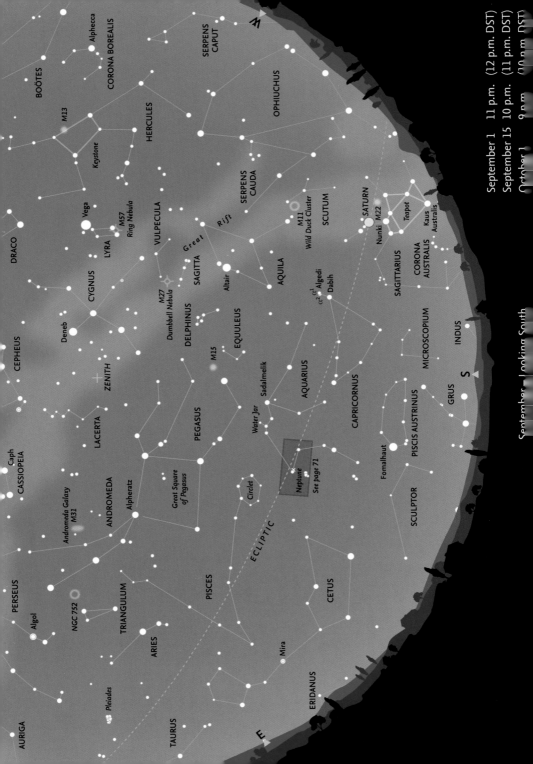

September 1   11 p.m.   (12 p.m. DST)
September 15   10 p.m.   (11 p.m. DST)
October 1   9 p.m.   (10 p.m. DST)

September — Looking South

September – Looking South

The **Summer Triangle** is now high in the southwest, with the Great Square of **Pegasus** high in the southeast. Below Pegasus are the two zodiacal constellations of **Capricornus** and **Aquarius.** In what is otherwise an unremarkable constellation, **Algedi** (α Capricorni) is actually a visual binary, with the two stars (α¹ Cap and α² Cap) readily seen with the naked eye. **Dabih** (β Capricorni), just to the south, is also a double star, and the components are relatively easy to separate with binoculars. In Aquarius, just to the east of **Sadalmelik** (α Aquari) there is a small asterism consisting of four stars, resembling a tiny letter "Y," known as the "Water Jar." Below Aquarius is a sparsely populated area of the sky with just one bright star in the constellation of **Piscis Austrinus.** In classical illustrations, water is shown flowing from the "Water Jar" toward bright **Fomalhaut** (α Piscis Austrini).

Another zodiacal constellation, **Pisces**, is now clearly visible to the east of Aquarius. Although faint, there is a distinctive asterism of stars, known as the "Circlet," south of the Great Square and another line of faint stars to the east of Pegasus. Still farther down toward the horizon

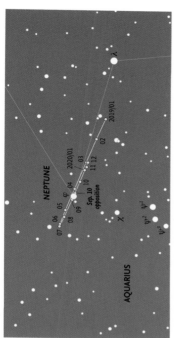

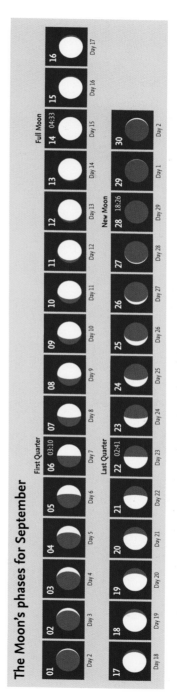

The path of Neptune in 2019. Neptune comes to opposition on September 10. All stars brighter than magnitude 8.5 are shown.

is the constellation of **Cetus**, with the famous variable star **Mira** (o Ceti) at its center. When Mira is at maximum brightness (around mag. 3.5) it is clearly visible to the naked eye, but it disappears as it fades toward minimum (about mag. 9.5 or less). There is a finder chart for Mira on page 83.

The Moon's phases for September

September – Moon and Planets

The Moon

On September 5, one day before First Quarter, the Moon is close to **Antares** and, the next day, near **Jupiter** in **Ophiuchus**. On September 8, 9 days old, it passes **Saturn** in **Sagittarius**. (An occultation is visible from Australia.) Full Moon (in **Pisces**) is on September 14. By September 20, as waning gibbous, the Moon is close to **Aldebaran** in **Taurus**. On September 26, the Moon is close to **Regulus** in **Leo**, visible in the early morning before sunrise. New Moon is on September 28 when the Moon (and the Sun) are in **Virgo**.

The planets

Mercury and **Venus** are too close to the Sun and too low to be visible. **Mars** is also lost in daylight. **Jupiter** (mag. -2.2 to -2.0) is moving eastwards slowly in **Ophiuchus**. **Saturn** (mag. 0.3 to 0.5) is also moving eastwards slowly in **Sagittarius**. **Uranus** (mag. 5.9) is in **Aries**. **Neptune** (in **Aquarius**) comes to opposition on September 10 at mag. 7.8 (on page 71 there is a finder chart showing the path of Neptune in 2019).

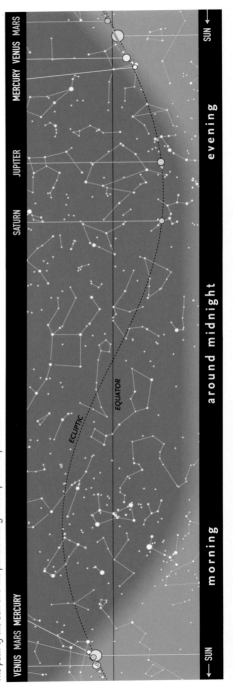

The path of the Sun and the planets along the ecliptic in September.

VENUS MARS MERCURY

morning around midnight evening

← SUN

SUN →

MERCURY VENUS MARS

JUPITER

SATURN

ECLIPTIC

EQUATOR

Calendar for September

02	09:13	Spica 7.7°S of Moon
02	10:42	Mars in conjunction with Sun
04	01:40	Mercury at superior conjunction
05	19:06	Antares 7.8°S of Moon
06	03:10	First Quarter
06	06:52	Jupiter 2.3°S of Moon
08	13:42	Saturn 0.1°N of Moon
10	07:24	Neptune at opposition (mag. 7.8)
13	13:32	Moon at apogee (406,377 km)
14		Full Moon
15	04:33	Alpha Aurigid shower second maximum
20	16:45	Aldebaran 2.7°S of Moon
22	02:41	Last Quarter
23	07:50	Autumnal equinox
23–Nov.19		Southern Taurid shower
23–Nov.27		Orionid meteor shower
24	00:01	Pollux 3.3°N of Moon
26	08:54	Regulus 3.3°S of Moon
28	01:19	Mars 4.1°S of Moon
28	02:24	Moon at perigee (357,802 km)
28	18:26	New Moon
29	12:46	Mars 4.4°S of Moon
29	19:44	Spica 7.6°S of Moon
29	22:01	Mercury 6.7°S of Moon

Evening 9:30 p.m. (DST)

September 5–8 • *The Moon passes Antares, Jupiter, Sabik, Nunki and Saturn.*

Midnight (DST)

September 19–20 • *The Moon with Aldebaran and the Pleiades.*

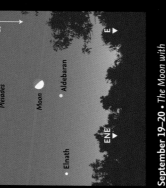

Early morning 3 a.m. (DST)

September 22–24 • *The Moon passes Alhena. On September 24, it lines up with Castor and Pollux*

Morning 6 a.m. (DST)

September 26 • *The Moon with Regulus and Algieba. Denebola is too close to the horizon to be readily seen*

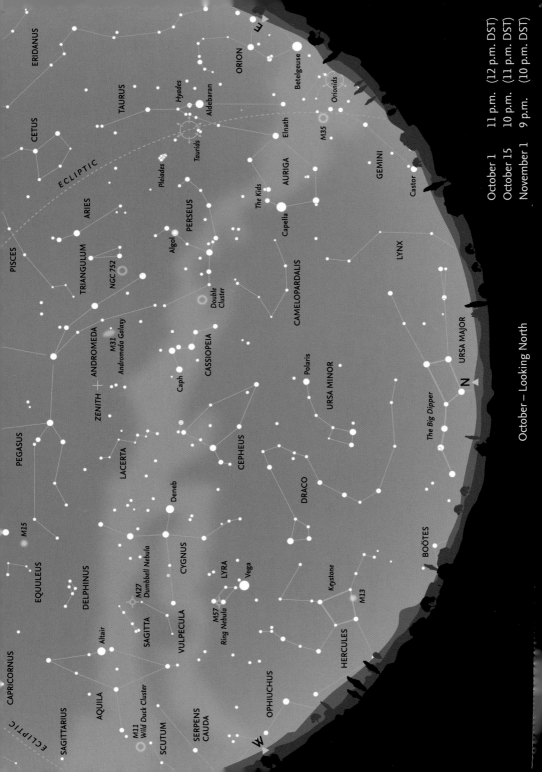

October – Looking North

October 1   11 p.m. (12 p.m. DST)
October 15   10 p.m. (11 p.m. DST)
November 1   9 p.m. (10 p.m. DST)

October – Looking North

Ursa Major is grazing the horizon in the north, while high overhead are the constellations of **Cepheus, Cassiopeia** and **Perseus**, with the Milky Way between Cepheus and Cassiopeia near the zenith. **Auriga** is now clearly visible in the east, as is **Taurus** with the **Pleiades, Hyades** and orange **Aldebaran**. Also in the east, **Orion** and **Gemini** are starting to rise clear of the horizon.

The constellations of **Boötes** and **Corona Borealis** are now lost to view in the northwest, and **Hercules** is also descending toward the western horizon. The three stars of the Summer Triangle are still clearly visible, although **Aquila** and **Altair** are beginning to approach the horizon in the west. Toward the end of the month (October 27) Summer Time ends in Europe. In North America the end of DST comes next month, in November.

Meteors

The **Orionids** are the major, fairly reliable meteor shower active in October. Like the May **Eta Aquariid** shower, the Orionids are associated with Comet 1P/Halley. During this second pass through the stream of particles from the comet, slightly fewer meteors are seen than in May, but conditions are more favourable for northern observers. In both showers the meteors are very fast, and many leave persistent trains. Although the Orionid maximum is quoted as October 21–22, in fact there is a very broad maximum, lasting about a week from October 20 to 27, with hourly rates around 25. Occasionally rates are higher (50–70 per hour). In 2019, the broad maximum extends from the day before Last Quarter to a waning crescent, so moonlight should not cause too much interference.

The faint shower of the **Southern Taurids** (often with bright fireballs) peaks on October 28–29. The Southern Taurid maximum occurs when the Moon is a waning crescent, so conditions are generally favourable. Towards the end of the month (around October 19), another shower (the **Northern Taurids**) begins to show activity, which peaks early in November. The parent comet for both Taurid showers is Comet 2P/Encke.

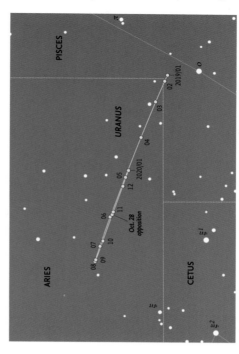

The path of Uranus in 2019. Uranus comes to opposition on October 28. All stars brighter than magnitude 7.5 are shown.

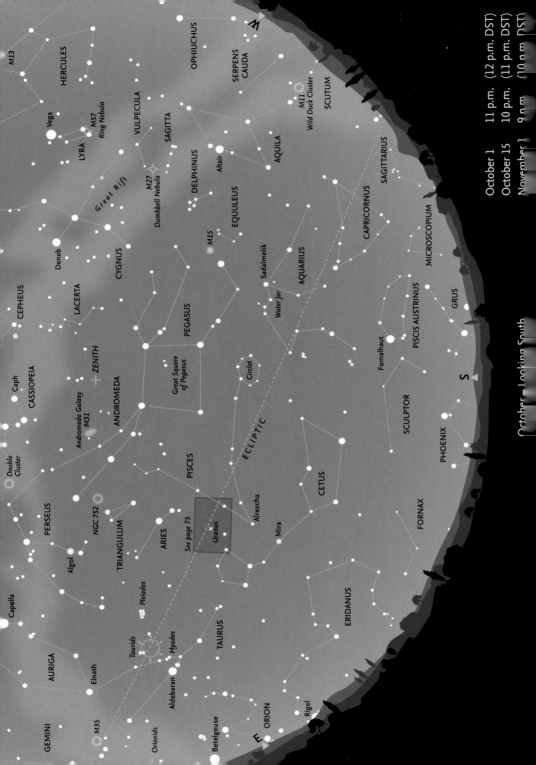

October 1   11 p.m.   (12 p.m. DST)
October 15   10 p.m.   (11 p.m. DST)
November 1   9 p.m.   (10 p.m. DST)

October Looking South

October – Looking South

The Great Square of **Pegasus** dominates the southern sky, framed by the two chains of stars that form the constellation of **Pisces**, together with **Alrescha** (α Piscium) at the point where the two lines of stars join. Also clearly visible is the constellation of **Cetus**, below Pegasus and Pisces. Although **Capricornus** is now lower, **Aquarius** to its east is well placed in the south, with solitary **Fomalhaut** and the constellation of **Piscis Austrinus** beneath it. The inconspicuous constellation of **Sculptor** appears in the south, as well as parts of **Grus**, **Phoenix** and, farther east, the northernmost stars of **Eridanus**.

The main band of the Milky Way and the Great Rift runs down from **Cygnus**, through **Vulpecula**, **Sagitta** and **Aquila** toward the western horizon. **Delphinus** and the tiny, unremarkable constellation of **Equuleus** lie between the band of the Milky Way and Pegasus. **Andromeda** is clearly visible high in the sky to the southeast, with the small constellation of **Triangulum** and the zodiacal constellation of **Aries** below it. **Perseus** is high in the east, and by now the **Pleiades** and **Taurus** are well clear of the horizon. Later in the night, and later in the month, **Orion** rises in the east, a sign that the autumn season has arrived and of the steady approach of winter.

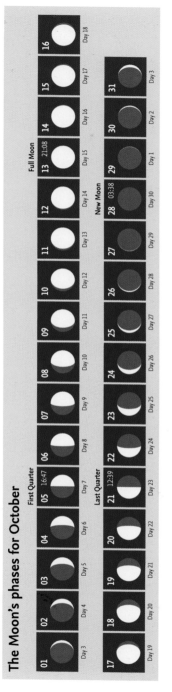

The constellation of Aquarius is one of the constellations that is visible in late summer and early autumn. The four stars forming the "Y"-shape of the "Water Jar" may be seen to the east of Sadalmelik (α Aquarii), the brightest star (top center).

The Moon's phases for October

				First Quarter						
01	02	03	04	05 16:47	06	07	08	09	10	11
Day 3	Day 4	Day 5	Day 6	Day 7	Day 8	Day 9	Day 10	Day 11	Day 12	Day 13

			Last Quarter							
12	13 21:08	14	15	16	17	18	19	20	21 12:39	22
Day 14	Day 15	Day 16	Day 17	Day 18	Day 19	Day 20	Day 21	Day 22	Day 23	Day 24

Full Moon

					New Moon					
23	24	25	26	27	28 03:38	29	30	31		
Day 25	Day 26	Day 27	Day 28	Day 29	Day 30	Day 1	Day 2	Day 3		

October – Moon and Planets

The Moon

The Moon is near *Antares* on October 3, but is visible only early just before setting. Later that day it is close to *Jupiter* and both bodies may be glimpsed just before they set in the west. On October 5, the Moon passes very close to *Saturn*. (There is an occultation visible from southern Africa.) Full Moon is on October 13. On October 17, the waning gibbous Moon is close to *Aldebaran* in *Taurus* and on October 23 it is near *Regulus* in *Leo*, but the constellation rises only later the next morning. New Moon occurs on October 28.

The planets

Mercury comes to greatest eastern elongation on October 20, but is invisible in daylight in *Libra*. *Venus* remains close to the Sun, also in Libra and not visible. *Mars* (in *Virgo*) is similarly too close to the Sun to be seen. *Jupiter* (mag. -2.0 to -1.9) is moving slowly eastwards in *Ophiuchus* and is visible in the early evening before it sets, as is *Saturn* in *Sagittarius* at mag. 0.5. *Uranus* remains in *Aries*, and comes to opposition on October 28 at mag. 5.7 (on page 75 there is a finder chart showing the path of Uranus in 2019). *Neptune* is in *Aquarius* at mag. 7.9. Both constellations are visible for a large part of the night.

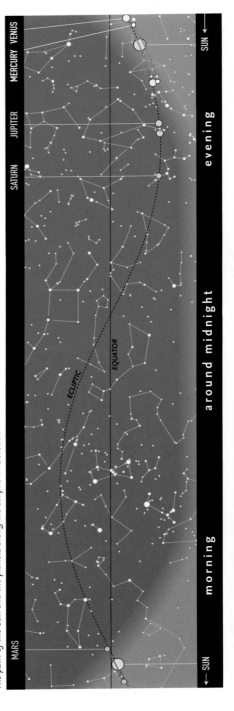

The path of the Sun and the planets along the ecliptic in October.

Calendar for October

03	01:00 *	Venus 3.1°N of Spica
03	03:18	Antares 7.5°S of Moon
03	20:23	Jupiter 1.9°S of Moon
05	16:47	First Quarter
05	20:36	Saturn 0.3°S of Moon (Occultation from S. Africa)
10	18:29	Moon at apogee (405,898 km)
13	21:08	Full Moon
17	22:22	Aldebaran 2.9°S of Moon
19–Dec.10		Northern Taurid meteor shower
20	04:02	Mercury at greatest elongation (24.6°E, mag. -0.1)
21	06:49	Pollux 5.6°N of Moon
21	12:39	Last Quarter
21–22		Orionid shower maximum
23	17:37	Regulus 3.5°S of Moon
26	10:39	Moon at perigee (363,101 km)
26	16:52	Mars 4.5°S of Moon
27	06:30	Spica 7.6°S of Moon
27		Summer Time ends (Europe)
28	03:38	New Moon
28	08:15	Uranus at opposition (mag. 5.7)
28–29		Southern Taurid shower maximum
29	14:55	Mercury 6.7°S of Moon
29	13:32	Venus 3.9°S of Moon
30	13:14	Antares 7.3°S of Moon
31	14:22	Jupiter 1.3°S of Moon

* These objects are close together for an extended period around this time.

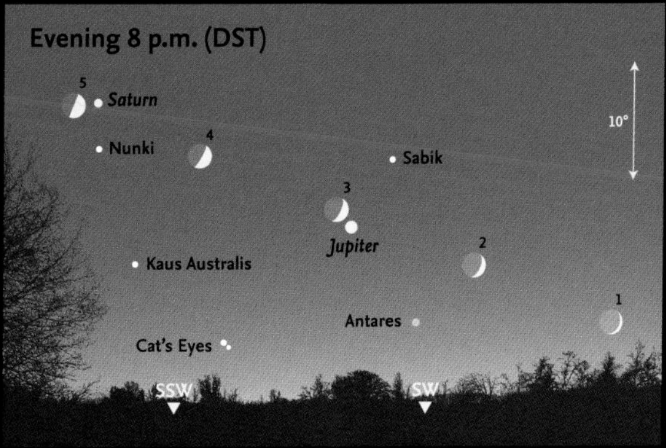

Evening 8 p.m. (DST)

October 1–5 • The Moon passes Antares, Sabik, Jupiter, Nunki and Saturn, in the southwestern sky.

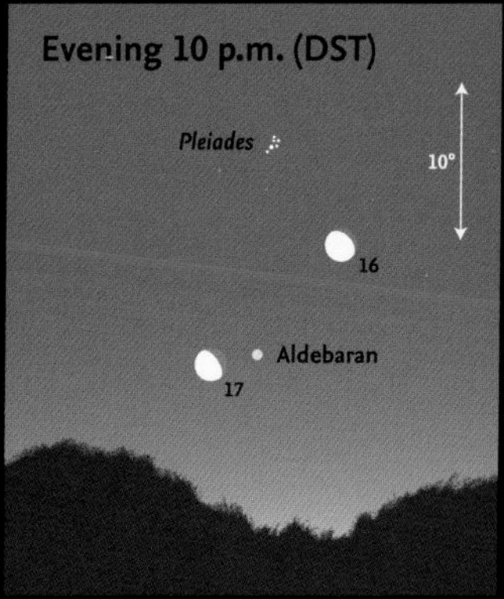

Evening 10 p.m. (DST)

October 17 • The Moon passes Aldebaran and the Pleiades.

Early morning 4 a.m. (DST)

October 23–24 • The Moon passes between Regulus and Algieba.

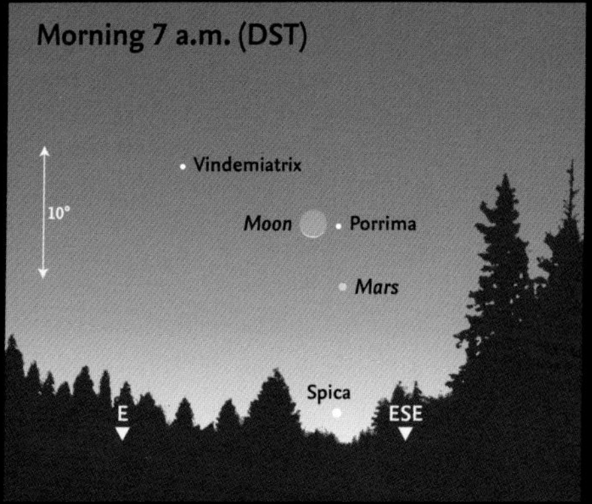

Morning 7 a.m. (DST)

October 27 • The Moon with Mars, Porrima (γ Vir) and Vindemiatrix (ε Vir). Spica is very low and lost in dawn.

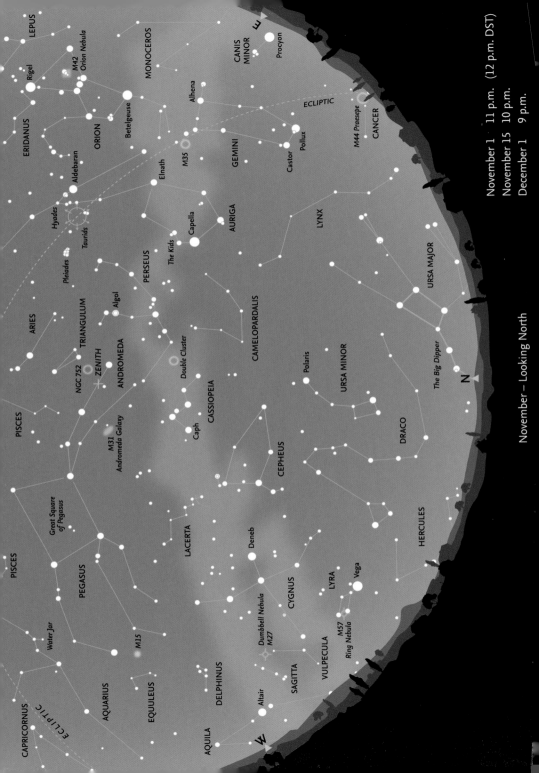

November 1   11 p.m.   (12 p.m. DST)
November 15   10 p.m.
December 1   9 p.m.

November – Looking North

November – Looking North

Most of **Aquila** has now disappeared below the horizon, but two of the stars of the Summer Triangle, **Vega** in **Lyra** and **Deneb** in **Cygnus**, are still clearly visible in the west. The head of **Draco** is now low in the northwest and only a small portion of **Hercules** remains above the horizon. The southernmost stars of **Ursa Major** are now coming into view. The Milky Way arches overhead, with the denser star clouds in the west and the less heavily populated region through **Auriga** and **Monoceros** in the east. High overhead, **Cassiopeia** is near the zenith and **Cepheus** has swung round to the northwest, while Auriga is now high in the northeast. **Gemini**, with **Castor** and **Pollux**, is well clear of the eastern horizon, and even **Procyon** (α Canis Minoris) is just climbing into view almost due east.

At the beginning of the month (November 3) Daylight Saving Time comes to an end in North America.

Meteors

The **Northern Taurid** shower, which began in mid-October, reaches maximum – although with only a low rate of about five meteors per hour – on November 12. Full Moon is on November 12, so conditions are extremely unfavorable. The shower gradually trails off, ending around December 10. There is an apparent 7-year periodicity in fireball activity, but 2019 is unlikely to be another peak year. Far more striking, however, are the **Leonids**, which have a short period of activity (November 5–30), with maximum on November 17–18. This shower is associated with Comet 55P/Tempel-Tuttle and has shown extraordinary activity on various occasions with many thousands of meteors per hour. High rates were seen in 1999, 2001 and 2002 (reaching about 3,000 meteors per hour) but have fallen dramatically since then. The rate in 2019 is likely to be about 15 per hour. These meteors are the fastest shower meteors recorded (about 70 km per second) and often leave persistent trains. The shower is very rich in faint meteors. In 2019, maximum is about a day before Last Quarter, when the Moon is waning gibbous, so conditions are not particularly favorable.

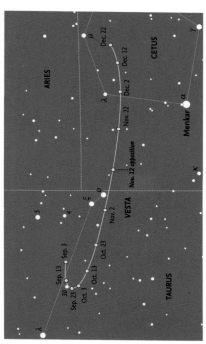

A finder chart for minor planet Vesta (4) which is at opposition (mag. 6.5) on November 12. Background stars are shown down to magnitude 7.5.

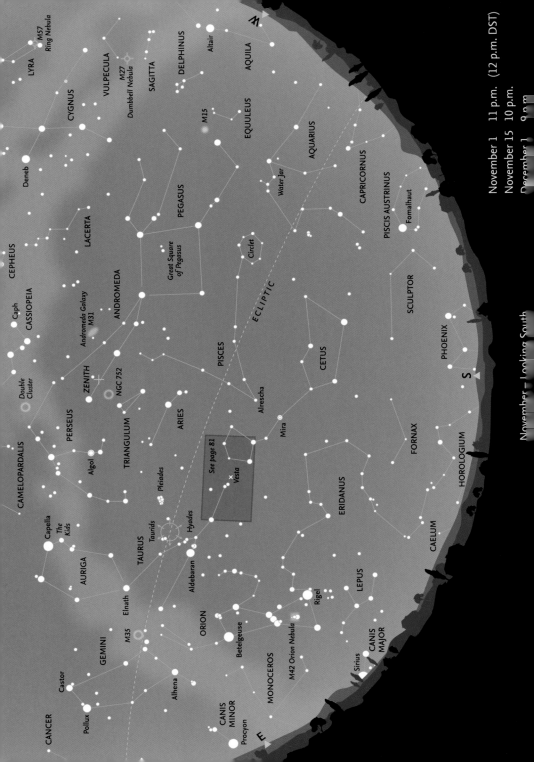

November — Looking South

November 1   11 p.m.   (12 p.m. DST)
November 15   10 p.m.
December 1   9 p.m.

November – Looking South

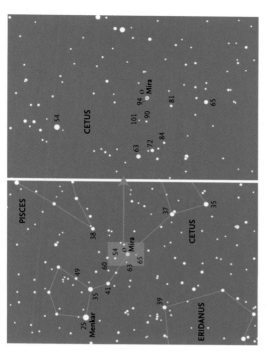

Orion has now risen above the eastern horizon, and part of the long, straggling constellation of **Eridanus** (which begins near **Rigel**) is visible to the west of Orion. Higher in the sky, **Taurus**, with the **Pleiades** cluster, and orange **Aldebaran** are now easy to observe. To their west, both **Pisces** and **Cetus** are close to the meridian. The famous long-period variable star, **Mira** (o Ceti), with a typical range of magnitude 3.4 to 9.8, is favorably placed for observation. In the southwest, **Capricornus** has slipped below the horizon, but **Aquarius** remains visible. Even farther west, **Altair** may be seen early in the night, but most of **Aquila** has already disappeared from view. **Delphinus**, together with **Sagitta** and **Vulpecula** in the Milky Way, will soon vanish for another year. Both **Pegasus** and **Andromeda** are easy to see, and one of the lines of stars that make up Andromeda passes close to the zenith.

Finder and comparison charts for Mira (o Ceti). The chart on the left shows all stars brighter than magnitude 6.5. The chart on the right shows stars down to magnitude 10.0. The comparison star magnitudes are shown without the decimal point.

The Moon's phases for November

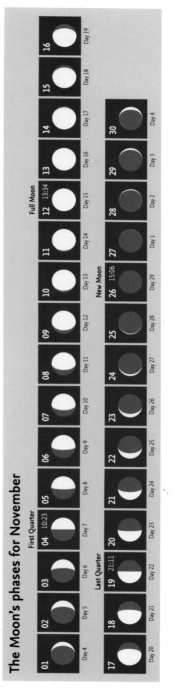

November – Moon and Planets

The Moon

On November 2 the Moon (a waxing crescent, two days before First Quarter) is very close to **Saturn** in **Sagittarius.** (There is an occultation, visible from New Zealand and the southern tip of Tasmania.) Full Moon is on November 12, when the Moon is on the border of **Aries** with **Taurus.** Two days later, on November 14, the Moon is north of **Aldebaran.** On November 19, at Last Quarter, it is near **Regulus** in **Leo.** On November 23, a waning crescent, it passes **Spica** in **Virgo** and, a day later, is close to **Mars.** The three bodies are visible in the early-morning sky. New Moon is on November 26, when it is in the small section of **Scorpius** between **Libra** and **Ophiuchus.** Two days later it is close to **Jupiter** in the morning sky, and on November 29 it is again close to **Saturn.**

The planets

Mercury is initially close to the Sun, passing inferior conjunction on November 11, but rapidly moves to greatest western elongation on November 28, when it may be glimpsed in the morning sky. **Venus** (mag. -3.8 to -3.9) is very low in the evening sky. **Mars** (mag. 1.8–1.7) is in **Libra,** visible in the early morning. **Jupiter** (mag. -1.9 to -1.8), initially in **Ophiuchus,** moves into **Sagittarius** and is visible in the southwest just before it sets. **Saturn** (mag. 0.5–0.6) is also in **Sagittarius. Uranus** is mag. 5.7 in **Aries** and **Neptune** (mag. 7.9) is in **Aquarius.** The minor planet (4) **Vesta** comes to opposition in **Cetus** on November 12. There is a finder chart on page 81.

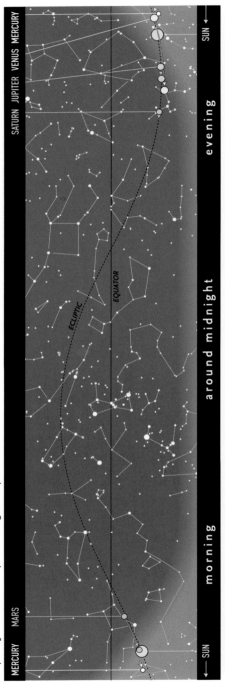

The path of the Sun and the planets along the ecliptic in November.

Calendar for November

02	07:21	Saturn 0.6°N of Moon
03		Daylight Saving Time ends in North America
04	10:23	First Quarter
05–30		Leonid meteor shower
07	08:36	Moon at apogee (405,058 km)
08	15:00 *	Mars 3.0°N of Spica
09	11:00 *	Venus 4.0°N of Antares
11	15:22	Mercury at inferior conjunction
12	08:56	Vesta at opposition (mag. 6.5)
12	13:34	Full Moon
14	04:23	Aldebaran 3.0°S of Moon
17	12:10	Pollux 5.4°N of Moon
17–18		Leonid shower maximum
19	21:11	Last Quarter
19	23:51	Regulus 3.7°S of Moon
23	07:41	Moon at perigee (366,716 km)
23	15:32	Spica 7.7°S of Moon
24	09:02	Mars 4.3°S of Moon
24	14:00 *	Venus 1.4°S of Jupiter
25	02:50	Mercury 1.9°S of Moon
26	15:06	New Moon
26	23:29	Antares 7.2°S of Moon
28	10:29	Mercury at greatest elongation (20.1°W, mag. -0.6)
28	10:49	Jupiter 0.7°S of Moon
28	18:49	Venus 1.9°S of Moon
29	21:03	Saturn 0.9°N of Moon

These objects are close together for an extended period around this time.

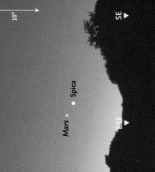

Morning 6 a.m.

November 9 • *Mars and Spica in the morning sky.*

Evening 5:15 p.m.

November 27–29 • *The Moon in the company of Jupiter, Venus and Saturn, low in the southwest.*

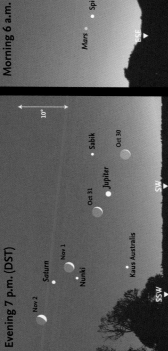

Evening 7 p.m. (DST)

October 30 – November 2 • *The Moon with Sabik, Jupiter, Nunki and Saturn, in the southwest, shortly after sunset.*

Morning 6:15 a.m.

November 23–25 • *The Moon passes Spica, Mars and Mercury in the southeast.*

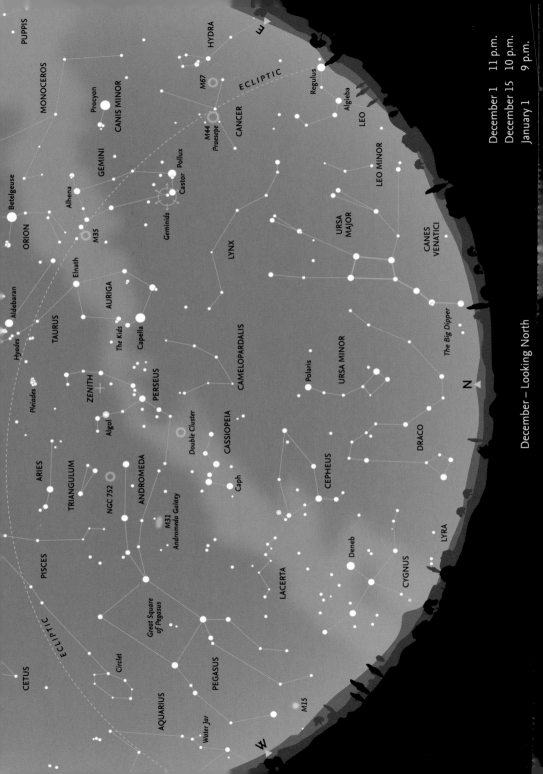

PUPPIS

MONOCEROS

HYDRA

CANIS MINOR

Procyon

M67

ECLIPTIC

Regulus

M44
Praesepe

CANCER

Algieba

LEO

GEMINI

Pollux

Castor

LEO MINOR

Alhena

Betelgeuse

Geminids

M35

LYNX

URSA MAJOR

ORION

Elnath

CANES VENATICI

AURIGA

TAURUS

The Kids

Capella

The Big Dipper

Aldebaran

Hyades

CAMELOPARDALIS

Pleiades

PERSEUS

Polaris

URSA MINOR

ZENITH

Algol

Double Cluster

ARIES

CASSIOPEIA

DRACO

N

TRIANGULUM

Caph

CEPHEUS

NGC 752

ANDROMEDA

M31
Andromeda Galaxy

PISCES

Deneb

Great Square
of Pegasus

LACERTA

LYRA

CETUS

Circlet

CYGNUS

PEGASUS

AQUARIUS

ECLIPTIC

Water Jar

M15

W

M15

December 1   11 p.m.
December 15   10 p.m.
January 1   9 p.m.

December – Looking North

December – Looking North

Ursa Major has now swung around and is starting to "climb" in the east. The fainter stars in the southern part of the constellation are now fully in view. The other bear, **Ursa Minor**, "hangs" below **Polaris** in the north. Directly above it is the faint constellation of **Camelopardalis**, with the other inconspicuous circumpolar constellation, **Lynx**, to its east. **Vega** (α Lyrae) is now below the horizon in the northwest, but **Deneb** (α Cygni) and most of **Cygnus** remain visible farther west. In the east, **Regulus** (α Leonis) and the constellation of **Leo** are beginning to rise above the horizon. **Cancer** stands high in the east, with **Gemini** even higher in the sky. **Perseus** is at the zenith, with **Auriga** and **Capella** between it and Gemini. Because it is so high in the sky, now is a good time to examine the star clouds of the fainter portion of the Milky Way, between **Cassiopeia** in the west to Gemini and **Orion** in the east.

Meteors

There is one significant meteor shower in December (the last major shower of the year). This is the **Geminid** shower, which is visible over the period December 4–16 and comes to maximum on December 13–14, just after Full Moon (December 12) so conditions are very poor. It is one of the most active showers of the year, and in some years is the strongest, with a peak rate of around 100 meteors per hour. It is the one major shower that shows good activity before midnight. The meteors (which are derived from cometary material). It was eventually established that the Geminids and the asteroid Phaethon had similar orbits. So the Geminids are assumed to consist of denser, rocky material. They are slower than most other meteors and often appear to last longer. The brightest often break up into numerous luminous fragments that follow similar paths across the sky. There is a second shower: the **Ursids**, active December 17–23, peaking on December

21–22, with rate at maximum of 5–10, occasionally rising to 25 per hour. Maximum in 2019 is when the Moon is a waning crescent, so conditions are reasonably favorable. The parent body is Comet 8P/Tuttle.

The constellation of Cassiopeia is a familiar sight among the northern circumpolar constellations. It is always above the horizon, on the opposite side of Polaris (the North Star) to the equally well-known asterism of the seven stars of the Big Dipper.

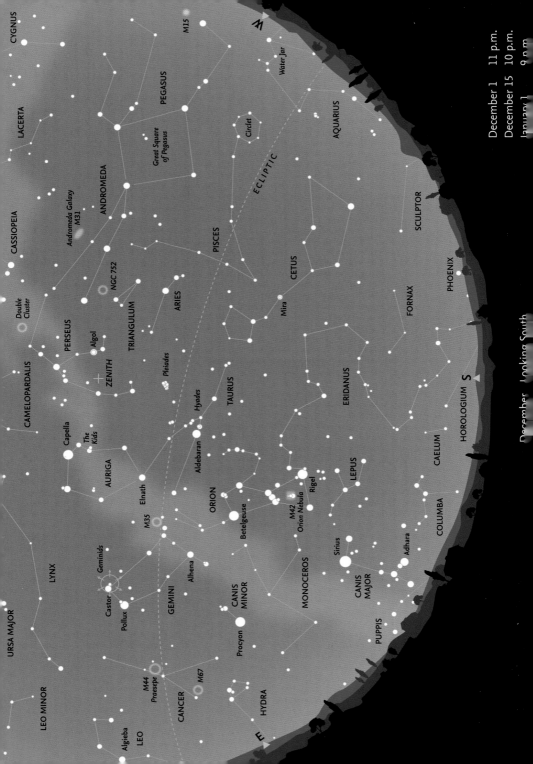

December 1   11 p.m.
December 15   10 p.m.
January 1   9 p.m.

December, Looking South

December – Looking South

The fine open cluster of the *Pleiades* is due south around 10 p.m., high in the sky, with the *Hyades* cluster, *Aldebaran* and the rest of *Taurus* clearly visible to the east. *Auriga* (with *Capella*) and *Gemini* (with *Castor* and *Pollux*) are both well placed for observation. *Orion* has made a welcome return to the winter sky, and both *Canis Minor* (with *Procyon*) and *Canis Major* (with *Sirius*, the brightest star in the sky) are now well above the horizon. The small, poorly known constellation of *Lepus* lies to the south of Orion, with *Columba* closer to the horizon. The northernmost portion of *Eridanus* is clearly visible in the south. In the west, *Aquarius* has now disappeared, and *Cetus* is becoming lower, but *Pisces* is still easily seen, as are the constellations of *Aries*, *Triangulum* and *Andromeda* above it. The Great Square of *Pegasus* is starting to plunge down toward the western horizon, and because of its orientation on the sky appears more like a large diamond, standing on one point, than a square.

The Moon's phases for December

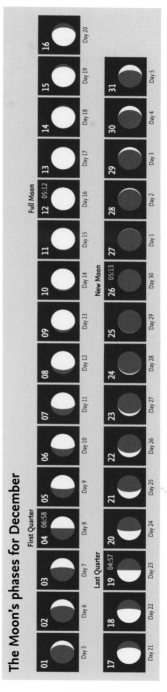

The constellation of Taurus contains two contrasting open clusters: the compact Pleiades, with its striking blue-white stars, and the more scattered, "V"-shaped Hyades, which are much closer to us. Orange Aldebaran (α Tauri) is not related to the Hyades, but lies between it and the Earth.

December – Moon and Planets

The Moon

The Moon is close to **Aldebaran** in **Taurus** on December 11, but Full Moon occurs the next day, so the star is overpowered by moonlight. On December 17, as waning gibbous, two days before Last Quarter, it is near **Regulus** in **Leo**, and on December 20, one day after Last Quarter, it is fairly close to **Spica** in **Virgo**. On December 23, as a waning crescent, it is close to **Mars** in **Libra**. At New Moon on December 26 there is an annular eclipse of the Sun, visible from Arabia, the Indian Ocean and Indonesia.

The planets

Mercury is rapidly approaching the Sun and is not visible. **Venus** begins the month low in the southwestern evening sky at mag. -3.9, and rapidly moves east into **Capricornus**, becoming more easily seen and brightening slightly. **Mars** (mag. 1.7) begins the month just inside **Virgo** and moves east into **Libra**. **Jupiter** is close to the Sun in **Sagittarius**, and **Saturn** is also in that constellation. **Uranus** is still just inside **Aries** near the border with **Pisces**, and **Neptune** remains in **Aquarius** where it has been throughout the year.

The path of the Sun and the planets along the ecliptic in December.

Evening 11 p.m.

Algieba
Regulus
Moon
ENE · E · 10°

December 16 • When the Moon rises in the east, it is between Regulus and Algieba.

Evening 5:30 p.m.

Venus
Moon
SW · 10°

December 29 • The Moon with Venus in the southwest, after sunset

Morning 5 a.m.

Elnath
Moon
Aldebaran
Pleiades
WNW · W · 10°

December 10–11 • The Moon is close to Aldebaran before it sets in the west.

Morning 6:45 a.m.

22
Mars
23
Antares
24
Sabik
ESE · SE · SSE · 10°

December 22–24 • The Moon passes Mars and Antares, just before sunrise. Antares and Sabik are probably lost in down

04	06:58	First Quarter
04–16		Geminid meteor shower
05	04:08	Moon at apogee (404,446 km)
.1	12:10	Aldebaran 3.0°S of Moon
2	05:12	Full Moon
3–14		Geminid shower maximum
.4	18:19	Pollux 5.3°N of Moon
7–23		Ursid meteor shower
.7	05:08	Regulus 3.8°S of Moon
8	20:25	Moon at perigee (370,265 km)
9	04:57	Last Quarter
0	22:03	Spica 7.8°S of Moon
1–22		Ursid shower maximum
2	04:19	Winter solstice
3	01:49	Mars 3.5°S of Moon
4	08:12	Antares 7.7°S of Moon
5	11:08	Mercury 1.9°S of Moon
6	05:13	New Moon
6	05:17	Annular solar eclipse (Arabia, Indian Ocean, Indonesia)
6	07:30	Jupiter 0.2°S of Moon
7	11:47	Saturn 1.2°N of Moon
7	18:26	Jupiter in conjunction with Sun
9	01:31	Venus 1.0°N of Moon

Glossary and Tables

aphelion	The point on an orbit that is farthest from the Sun.
apogee	The point on its orbit at which the Moon is farthest from the Earth.
appulse	The apparently close approach of two celestial objects; two planets, or a planet and star.
astronomical unit	(AU) The mean distance of the Earth from the Sun, 149,597,870 km.
celestial equator	The great circle on the celestial sphere that is in the same plane as the Earth's equator.
celestial sphere	The apparent sphere surrounding the Earth on which all celestial bodies (stars, planets, etc.) seem to be located.
conjunction	The point in time when two celestial objects have the same celestial longitude. In the case of the Sun and a planet, superior conjunction occurs when the planet lies on the far side of the Sun (as seen from Earth). For Mercury and Venus, inferior conjuction occurs when they pass between the Sun and the Earth.
direct motion	Motion from west to east on the sky.
ecliptic	The apparent path of the Sun across the sky throughout the year. Also: the plane of the Earth's orbit in space.
elongation	The point at which an inferior planet has the greatest angular distance from the Sun, as seen from Earth.
equinox	The two points during the year when night and day have equal duration. Also: the points on the sky at which the ecliptic intersects the celestial equator. The vernal (spring) equinox is of particular importance in astronomy.
gibbous	The stage in the sequence of phases at which the illumination of a body lies between half and full. In the case of the Moon, the term is applied to phases between First Quarter and Full, and between Full and Last Quarter.
inferior planet	Either of the planets Mercury or Venus, which have orbits inside that of the Earth.
magnitude	The brightness of a star, planet or other celestial body. It is a logarithmic scale, where larger numbers indicate fainter brightness. A difference of 5 in magnitude indicates a difference of 100 in actual brightness, thus a first-magnitude star is 100 times as bright as one of sixth magnitude.
meridian	The great circle passing through the North and South Poles of a body and the observer's position; or the corresponding great circle on the celestial sphere that passes through the North and South Celestial Poles and also through the observer's zenith.
nadir	The point on the celestial sphere directly beneath the observer's feet, opposite the zenith.
occultation	The disappearance of one celestial body behind another, such as when stars or planets are hidden behind the Moon.
opposition	The point on a superior planet's orbit at which it is directly opposite the Sun in the sky.
perigee	The point on its orbit at which the Moon is closest to the Earth.
perihelion	The point on an orbit that is closest to the Sun.
retrograde motion	Motion from east to west on the sky.
superior planet	A planet that has an orbit outside that of the Earth.
vernal equinox	The point at which the Sun, in its apparent motion along the ecliptic, crosses the celestial equator from south to north. Also known as the First Point of Aries.
zenith	The point directly above the observer's head.
zodiac	A band, stretching 8° on either side of the ecliptic, within which the Moon and planets appear to move. It consists of 12 equal areas, originally named after the constellation that once lay within it.

The Constellations

There are 88 constellations covering the whole of the celestial sphere, but 16 of these in the southern hemisphere can never be seen (even in part) from a latitude of 40°N, so are omitted from this table. The names themselves are expressed in Latin, and the names of stars are frequently given by Greek letters (see next page) followed by the genitive of the constellation name. The genitives and English names of the various constellations are included.

Name	Genitive	Abbr.	English name
Andromeda	Andromeda	And	Andromeda
Antlia	Antliae	Ant	Air Pump
Aquarius	Aquarii	Aqr	Water Bearer
Aquila	Aquilae	Aql	Eagle
Ara	Arae	Ara	Altar
Aries	Arietis	Ari	Ram
Auriga	Aurigae	Aur	Charioteer
Boötes	Boötis	Boo	Herdsman
Caelum	Caeli	Cae	Burin (Chisel)
Camelopardalis	Camelopardalis	Cam	Giraffe
Cancer	Cancri	Cnc	Crab
Canes Venatici	Canum Venaticorum	CVn	Hunting Dogs
Canis Major	Canis Majoris	CMa	Big Dog
Canis Minor	Canis Minoris	CMi	Little Dog
Capricornus	Capricorni	Cap	Sea Goat
Cassiopeia	Cassiopeiae	Cas	Cassiopeia
Centaurus	Centauri	Cen	Centaur
Cepheus	Cephei	Cep	Cepheus
Cetus	Ceti	Cet	Whale
Columba	Columbae	Col	Dove
Coma Berenices	Comae Berenices	Com	Berenice's Hair
Corona Australis	Coronae Australis	CrA	Southern Crown
Corona Borealis	Coronae Borealis	CrB	Northern Crown
Corvus	Corvi	Crv	Crow
Crater	Crateris	Crt	Cup
Cygnus	Cygni	Cyg	Swan
Delphinus	Delphini	Del	Dolphin
Draco	Draconis	Dra	Dragon
Equuleus	Equulei	Equ	Little Horse
Eridanus	Eridani	Eri	River Eridanus
Fornax	Fornacis	For	Furnace
Gemini	Geminorum	Gem	Twins
Grus	Gruis	Gru	Crane
Hercules	Herculis	Her	Hercules
Horologium	Horologii	Hor	(Pendulum) Clock
Hydra	Hydrae	Hya	Water Snake

Name	Genitive	Abbr.	English name
Indus	Indi	Ind	Indian
Lacerta	Lacertae	Lac	Lizard
Leo	Leonis	Leo	Lion
Leo Minor	Leonis Minoris	LMi	Little Lion
Lepus	Leporis	Lep	Hare
Libra	Librae	Lib	Scales
Lupus	Lupi	Lup	Wolf
Lynx	Lyncis	Lyn	Lynx
Lyra	Lyrae	Lyr	Lyre
Microscopium	Microscopii	Mic	Microscope
Monoceros	Monocerotis	Mon	Unicorn
Norma	Normae	Nor	Level (Square)
Ophiuchus	Ophiuchi	Oph	Serpent Bearer
Orion	Orionis	Ori	Orion
Pegasus	Pegasi	Peg	Pegasus
Perseus	Persei	Per	Perseus
Phoenix	Phoenicis	Phe	Phoenix
Pisces	Piscium	Psc	Fishes
Piscis Austrinus	Piscis Austrini	PsA	Southern Fish
Puppis	Puppis	Pup	Stern
Pyxis	Pyxidis	Pyx	Compass
Sagitta	Sagittae	Sge	Arrow
Sagittarius	Sagittarii	Sgr	Archer
Scorpius	Scorpii	Sco	Scorpion
Sculptor	Sculptoris	Scl	Sculptor
Scutum	Scuti	Sct	Shield
Serpens	Serpentis	Ser	Serpent
Sextans	Sextantis	Sex	Sextant
Taurus	Tauri	Tau	Bull
Telescopium	Telescopii	Tel	Telescope
Triangulum	Trianguli	Tri	Triangle
Ursa Major	Ursae Majoris	UMa	Great Bear
Ursa Minor	Ursae Minoris	UMi	Lesser Bear
Vela	Velorum	Vel	Sails
Virgo	Virginis	Vir	Virgin
Vulpecula	Vulpeculae	Vul	Fox

The Greek Alphabet

α	Alpha	ε	Epsilon	ι	Iota	ν	Nu	ρ	Rho
β	Beta	ζ	Zeta	κ	Kappa	ξ	Xi	σ (ς)	Sigma
γ	Gamma	η	Eta	λ	Lambda	o	Omicron	τ	Tau
δ	Delta	θ (ϑ)	Theta	μ	Mu	π	Pi	υ	Upsilon

φ (φ)	Phi
χ	Chi
ψ	Psi
ω	Omega

Some common asterisms

Belt of Orion	δ, ε, and ζ Orionis
Big Dipper	α, β, γ, δ, ε, ζ and η Ursae Majoris
Cat's Eyes	λ and υ Scorpii
Circlet	γ, θ, ι, λ and κ Piscium
Guards (or Guardians)	β and γ Ursae Minoris
Head of Cetus	α, γ, ξ², μ and λ Ceti
Head of Draco	β, γ, ξ and ν Draconis
Head of Hydra	δ, ε, ζ, η, ρ and σ Hydrae
Keystone	ε, ζ, η and π Herculis
Kids	ε, ζ and η Aurigae
Little Dipper	β, γ, η, ζ, ε, δ and α Ursae Minoris
Lozenge	= Head of Draco
Milk Dipper	ζ, γ, σ, φ and λ Sagittarii
Plough or Big Dipper	α, β, γ, δ, ε, ζ and η Ursae Majoris
Pointers	α and β Ursae Majoris
Sickle	α, η, γ, ζ, μ and ε Leonis
Square of Pegasus	α, β and γ Pegasi with α Andromedae
Sword of Orion	θ and ι Orionis
Teapot	γ, ε, δ, λ, φ, σ, τ and ζ Sagittarii
Wain (or Charles' Wain)	= Big Dipper
Water Jar	γ, η, κ and ζ Aquarii
Y of Aquarius	= Water Jar

Acknowledgments

Damian Peach, Hamble, Hants.: p.12 (Comet Lovejoy)
Sjbmgrtl: p.13 (Comet McNaught)
[https://commons.wikimedia.org/wiki/File.Sat_comet_WEB.jpg]
peresanz/Shutterstock: p.21 (Orion)
Steve Edberg, La Cañada, California: all other constellation photographs
Denis Buczynski, Portmahomack, Ross-shire, p.23 (Quadrantid fireball); p.51 (noctilucent clouds)
Jens Hackmann: p.63 (Perseid meteor)
Ken Sperber, California: p.69 (Double Cluster)
Steve Edberg, La Cañada, California: all other constellation photographs

Specialist editorial support was provided by Gregory Brown, Astronomy Education Officer at the Royal Observatory Greenwich, part of Royal Museums Greenwich.

Further Information

Books

Bone, Neil (1993), *Observer's Handbook: Meteors*, George Philip, London & Sky Publ. Corp., Cambridge, Mass.

Cook, J., ed. (1999), *The Hatfield Photographic Lunar Atlas*, Springer-Verlag, New York

Dunlop, Storm (2006), *Wild Guide to the Night Sky*, Harper Perennial, New York & Smithsonian Press, Washington D.C.

Dunlop, Storm (2012), *Practical Astronomy*, 2nd edn, Firefly, Buffalo

Dunlop, Storm, Rükl, Antonin & Tirion, Wil (2005), *Collins Atlas of the Night Sky*, HarperCollins, London & Smithsonian Press, Washington D.C.

Grego, Peter (2016), *Moon Observer's Guide*, Firefly, Richmond Hill

Heifetz, Milton D. & Tirion, Wil (2017), *A Walk through the Heavens*, 4th edn, Cambridge University Press, Cambridge

Mellinger, Axel & Hoffmann, Susanne (2005), *The New Atlas of the Stars*, Firefly, Richmond Hill

O'Meara, Stephen J. (2008), *Observing the Night Sky with Binoculars*, Cambridge University Press, Cambridge

Pasachoff, Jay M. (1999), *Peterson Field Guides: Stars and Planets*, 4th edn., Houghton Mifflin, Boston

Ridpath, Ian, ed. (2004), *Norton's Star Atlas*, 20th edn, Pi Press, New York

Ridpath, Ian, ed. (2003), *Oxford Dictionary of Astronomy*, 2nd edn, Oxford University Press, Oxford & New York

Ridpath, Ian & Tirion, Wil (2004), *Collins Gem – Stars*, HarperCollins, London

Ridpath, Ian & Tirion, Wil (2011), *Collins Pocket Guide Stars and Planets*, 4th edn, HarperCollins, London

Ridpath, Ian & Tirion, Wil (2012), *Monthly Sky Guide*, 9th edn, Cambridge University Press, Cambridge

Rükl, Antonín (1990), *Hamlyn Atlas of the Moon*, Hamlyn, London & Astro Media Inc., Milwaukee

Rükl, Antonín (2004), *Atlas of the Moon*, Sky Publishing Corp., Cambridge, Mass.

Scagell, Robin (2015), *Firefly Complete Guide to Stargazing*, Firefly, Richmond Hill

Scagell, Robin & Frydman, David (2014), *Stargazing with Binoculars*, Firefly, Richmond Hill

Scagell, Robin (2014), *Stargazing with a Telescope*, Firefly, Richmond Hill

Sky & Telescope (2017), *Astronomy 2018*, Sky Publishing Corp., Cambridge, Mass.

Tirion, Wil (2011), *Cambridge Star Atlas*, 4th edn, Cambridge University Press, Cambridge

Tirion, Wil & Sinnott, Roger (1999), *Sky Atlas 2000.0*, 2nd edn, Sky Publishing Corp., Cambridge, Mass. & Cambridge University Press, Cambridge

Journals

Astronomy, Astro Media Corp., 21027 Crossroads Circle, P.O. Box 1612, Waukesha, WI 53187-1612.
http://www.astronomy.com

Sky & Telescope, Sky Publishing Corp., Cambridge, MA 02138-1200.
http://www.skyandtelescope.com/

Societies

American Association of Variable Star Observers (AAVSO), 49 Bay State Rd., Cambridge, MA 02138.
Although primarily concerned with variable stars, the AAVSO also has a solar section.

American Meteor Society (AMS), Geneseo, New York.
http://www.amsmeteors.org/

Association of Lunar and Planetary Observers (ALPO), ALPO Membership Secretary/Treasurer, P.O. Box 13456, Springfield, IL 62791-3456.
An organization concerned with all forms of amateur astronomical observation, not just the Moon and planets, with numerous coordinated observing sections.
http://alpo-astronomy.org/

Astronomical League (AL), 9201 Ward Parkway Suite #100, Kansas City, MO 64114.
An umbrella organization consisting of over 240 local amateur astronomical societies across the United States.
https://www.astroleague.org/

British Astronomical Association (BAA), Burlington House, Piccadilly, London W1J 0DU.

> The principal British organization (but with a worldwide membership) for amateur astronomers (with some professional members), particularly for those interested in carrying out observational programs.
> http://www.britastro.org/

International Meteor Organization (IMO)

> An organization coordinating observations of meteors worldwide.
> http://www.imo.net/

Royal Astronomical Society of Canada (RASC), 203 – 4920 Dundas Street W., Toronto, ON M9A 1B7.

> The principal Canadian astronomical organization, with both professional and amateur members. It has 28 local centers.
> http://rasc.ca/

Software

Planetary, Stellar and Lunar Visibility (planetary and eclipse freeware): Alcyone Software, Germany.
> http://www.alcyone.de

Redshift, Redshift-Live.
> http://www.redshift-live.com/en/

Starry Night & Starry Night Pro, Sienna Software Inc., Toronto, Canada.
> http://www.starrynight.com

Internet sources

There are numerous sites about all aspects of astronomy, and all have numerous links. Although many amateur sites are excellent, treat any statements and data with caution. The sites listed below offer accurate information. Please note that the URLs may change. If so, use a good search engine, such as Google, to locate the information source.

Information

Astronomical data (inc. eclipses) HM Nautical Almanac Office: http://astro.ukho.gov.uk

Auroral information Michigan Tech: http://www.geo.mtu.edu/weather/aurora/

Comets JPL Solar System Dynamics: http://ssd.jpl.nasa.gov/

Deep-sky objects Saguaro Astronomy Club Database: http://www.virtualcolony.com/sac/

Eclipses NASA Eclipse Page: http://eclipse.gsfc.nasa.gov/eclipse.html

Moon (inc. Atlas) Inconstant Moon: http://www.inconstantmoon.com/

Planets Planetary Fact Sheets: http://nssdc.gsfc.nasa.gov/planetary/planetfact.html

Satellites (inc. International Space Station)

> Heavens Above: http://www.heavens-above.com/
> Visual Satellite Observer: http://www.satobs.org/

Star Chart http://www.skyandtelescope.com/observing/interactive-sky-watching-tools/interactive-sky-chart/

What's Visible

> Skyhound: http://www.skyhound.com/sh/skyhound.html
> Skyview Cafe: http://www.skyviewcafe.com

Institutes and Organizations

European Space Agency: http://www.esa.int/

International Dark-Sky Association: http://www.darksky.org/

Jet Propulsion Laboratory: http://www.jpl.nasa.gov/

Lunar and Planetary Institute: http://www.lpi.usra.edu/

National Aeronautics and Space Administration: http://www.hq.nasa.gov/

Solar Data Analysis Center: http://umbra.gsfc.nasa.gov/

Space Telescope Science Institute: http://www.stsci.edu/